SPC Essentials and Productivity Improvement

Also available from ASQ Quality Press

Transparency Master to Accompany
SPC Essentials and Productivity Improvement: A Manufacturing
Approach
William A. Levinson and Frank Tumbelty

Statistical Process Control Methods for Long and Short Runs, Second Edition
Gary K. Griffith

Glossary and Tables for Statistical Quality Control, Third Edition
ASQ Statistics Division

SPC Tools for Everyone
John T. Burr

Quality Control for Operators and Foremen
K. S. Krishnamoorthi

Concepts for R&R Studies
Larry B. Barrentine

To request a complimentary catalog of publications, call 800-248-1946.

SPC Essentials and Productivity Improvement

A Manufacturing Approach

William A. Levinson
Frank Tumbelty
Harris Corporation, Semiconductor Sector

ASQ Quality Press
Milwaukee, Wisconsin

SPC Essentials and Productivity Improvement: A Manufacturing Approach
William A. Levinson and Frank Tumbelty

Library of Congress Cataloging-in-Publication Data
Levinson, William A., 1957–
    SPC essentials and productivity improvement: a manufacturing
approach / William A. Levinson, Frank Tumbelty.
        p.   cm.
    Includes bibliographical references and index.
    ISBN 0-87389-372-7 (alk. paper)
    1. Process control—Statistical methods.   2. Production
management.   I. Tumbelty, Frank, 1934–   . II. Title.
TS156.8.L48   1997
658.5'015195—dc20   96-34021
                                                              CIP

Many of the designations used by manufacturers and sellers to distinguish their products
are claimed as trademarks. Where those designations appear in this book and ASQ Quality
Press was aware of a trademark claim, the designations have been printed in initial caps.

10 9 8 7 6 5 4 3

ISBN 0-87389-372-7

Acquisitions Editor: Roger Holloway
Project Editor: Jeanne W. Bohn

ASQ Mission: To facilitate continuous improvement and increase customer satisfaction
by identifying, communicating, and promoting the use of quality principles, concepts,
and technologies; and thereby be recognized throughout the world as the leading
authority on, and champion for, quality.

Attention: Schools and Corporations
ASQ Quality Press books, audiotapes, videotapes, and software are available at quantity
discounts with bulk purchases for business, educational, or instructional use. For information, please contact ASQ Quality Press at 800-248-1946, or write to ASQ Quality
Press, P.O. Box 3005, Milwaukee, WI 53201-3005.

For a free copy of the ASQ Quality Press Publications Catalog, including ASQ mem-
bership information, call 800-248-1946.

Printed in the United States of America

∞ Printed on acid-free paper

American Society for Quality

ASQ

Quality Press
611 East Wisconsin Avenue
Milwaukee, Wisconsin 53201-3005
800-248-1946
Web site http://www.asq.org

To my father, Herbert H. Levinson.

W. A. L.

To my wife, Lynn.

F. T.

Contents

Preface

Quality improvement methods exist to support manufacturing operations and to make them more effective. To be useful on the factory floor, the methods must be practical and easy to use. Robert A. Heinlein's *Starship Troopers* expands on this concept. If your equipment is complex and hard to use, someone will hit you over the head with something simple—like a stone ax—while you are trying to read a vernier.*

This does not mean we want to use primitive methods because they are simple. In Heinlein's story, soldiers wear sophisticated, mechanically powered suits of armor. The armor automatically follows its wearer's motions, so it is easy to use. It's like wearing an overcoat. We will take this approach with quality improvement techniques for the factory. Computers will handle the complex math and deliver charts and reports that are easy to use.

The material in this book uses basic math: addition, subtraction, multiplication, and division. In industry, time costs money. As a practical process management technique, statistical process control (SPC) does not require lengthy manual calculations. If the math is not simple enough for rapid calculation by hand, a computer handles it.

We must, however, understand what the computer is telling us about the process. Chapter 3 equips readers to understand and interpret the control charts they may see in the factory. Chapter 2 covers nonmathematical techniques for improving quality and productivity, and quality management techniques. It answers questions like, "Why should we use ISO 9000 standards?"

A basic SPC and quality improvement course should use chapters 1 through 5, but omit the technical appendices for chapters 4 and 5. An advanced course should add the technical appendices and chapter 6. These

---

*Technology has surpassed even the imaginations of twentieth century science fiction writers. Many modern gages measure pieces automatically, so some readers may be unfamiliar with verniers. Verniers are mechanical aids for interpolating a gage or instrument. Another Heinlein story from the 1950s has astronauts on an intersteller spaceship using slide rules. Readers under 30 may be unfamiliar with slide rules, which are now seen only in museums. Slide rules are mechanical computing devices that people used before there were calculators.

sections equip readers to program a computer or tell commercial SPC software what to do. They show how to characterize processes and set up new control charts for them. This material requires technical mathematics at the advanced high school or freshman college level.

As manufacturing workers, technicians, or engineers, readers play vital roles in organizational competitiveness. We will start by discussing the role of manufacturing and quality in national wealth, security, and prosperity.

CHAPTER ONE

Introduction

Manufacturing, or adding value to physical products, is the keystone of national wealth and prosperity. Quality helps us sell our products and stay in business and helps us take our competitors' business. Quality means *fitness for use*, and the customer defines it. This section discusses qualitative and quantitative measures for assuring fitness for use. Manufacturing workers have more familiarity with the job than anyone else. They play a principal role in assuring and improving productivity and quality.

The Importance of Manufacturing

Physical wealth comes only from natural resources and manufacturing. Manufacturing adds value to materials or intermediate parts, such as subassemblies or work in process. We will treat construction like manufacturing, since turning materials into a house, ship, or bridge makes them more valuable. Manufacturing is the primary source of national wealth.

Workers in extractive industries harvest natural resources and turn them into raw materials. Examples of these industries include lumber; mining (isolating metals from their ores); petroleum (getting the petroleum out of the ground and turning it into gasoline and other chemicals); fishing; and agriculture.

Except for agricultural products and fish, natural resources are of little use by themselves. For example, silicon is the second most common element in the earth's crust. A handful of sand on the beach is 47 percent silicon. We wouldn't expect it to be valuable like gold, which is scarce. Semiconductor manufacturers buy silicon and turn it into computer chips, power controls, and other electronic products. If King Midas turned a semiconductor factory's finished goods into gold, he'd destroy most of their value!

We can compare manufacturing to medieval alchemy. The alchemists wanted to turn base materials, like lead, into gold. The microelectronics industry turns sand into tiny electronic parts that are more valuable than gold.

The word *manufacture* actually means *to make by hand*. The Latin word for *hand* gave the English language *manual*, *manipulate*, and *manufacture*. The second syllable *facture* comes from the Latin word for *to make*, and this word is the ancestor of *factory*. In Roman times, no one had automatic machinery, so all production work was by hand. A blacksmith would take iron (a raw material) and use hand tools to turn it into something useful. A weaver would take cotton, an agricultural product, and work a hand loom to turn the cotton into cloth. Carpenters used hand saws and hammers to shape wood into furniture and other items. These activities added value to the raw materials. A plowshare was more valuable than an ingot of pig iron, and a chair was more valuable than a log.

No other industries create physical wealth. These include the following:

- Transportation (trucking, shipping, railroads, airplanes) moves raw materials or products, but does not create them.

- Financial services (banking, stock trading) allow people to lend, borrow, or invest wealth, but do not create it.

- Distribution (retailing, wholesaling) delivers products to customers, but does not create any wealth. Providers of goods or services are trying to cut intermediaries out of the distribution chain. Mail order catalogs, including electronic ones on the Internet, are taking business from retailers. The catalog retailer avoids the costs of renting a store and hiring people to staff it.

This does not mean that transportation, financial services, and distribution are not useful. All of them play supporting roles in creating and distributing wealth. People use financial services to sell stock or borrow money to build factories. Transportation services deliver raw materials to factories and take finished goods to customers.

Retailers and wholesalers provide a convenient, central marketplace for goods. Without them, we'd have to make separate trips to the baker, butcher, dairy, fishery, and so on. The cost of doing this would far exceed the retailer's markup. Meanwhile, there is no practical way to do all of one's grocery shopping via mail order. Some goods are perishable, while others are bulky, heavy, and costly to ship. Some imaginative grocers offer computerized shopping and local delivery services, though.

Albert W. Moore (1994), president of the Association for Manufacturing Technology (AMT), provides an excellent summary: "Manufacturing and the export of manufactured products is the only way to guarantee our national security, international strength, and future prosperity."

Manufacturing and Wealth: A Historical Lesson

National power, whether economic or military, comes from manufacturing capability: the ability to add value to physical products. There is a cynical version of the Golden Rule that says, "The person with the gold makes the rules." Actually, the people with the gold don't make the rules. The people with the factories make the rules.

During the fifteenth and early sixteenth centuries, Spaniards and Portuguese were the world's foremost navigators and explorers. Vasco da Gama sailed to India, and King Ferdinand and Queen Isabella financed Christopher Columbus' explorations. Ferdinand Magellan, a Portuguese explorer, discovered the passage around the tip of South America, but Spanish and Portuguese maritime dominance ended within a century of Columbus' voyage. The discovery of gold in the New World actually handed this dominance to England and Holland.

The English and Dutch did not get any gold from the New World, but their North American colonies shipped raw materials such as cotton back to Europe. English and Dutch factories converted the raw materials into useful products. Gold made it easy for Spain and Portugal to buy manufactured goods from England and Holland. This demand encouraged the British and Dutch to build factories to make the goods and the ships to transport them. This helped the British and Dutch shipbuilding industries, at the expense of the Spanish and Portuguese. Spanish shipbuilding declined because it did not take many ships to transport gold. Eventually, Spain produced little more than wool, fruit, and iron—all products of agriculture and extractive industries.

By 1648, Spain's merchant marine was in such contemptible condition that Spain had to hire Dutch ships to service the Indies. This was despite Spain's (the Inquisition's) view that the Protestant Dutch were heretics. The Spanish Navy was not a major factor in world politics after its defeat at Gravelines (the Armada, 1588). The seventeenth century saw several battles between England and Holland for maritime supremacy. England finally triumphed and became the most powerful country on earth. A famous American naval captain, Alfred Thayer Mahan (1980), explained what happened. "The tendency to trade, *involving of necessity the production of something to trade with* [author's emphasis], is the national characteristic most important to the development of sea power" (p. 46).

During the eighteenth and nineteenth centuries, Great Britain imported raw materials from its colonies and processed them into finished goods. The Navigation Acts mandated that English ships had to carry all commerce that went through English ports.* This helped English shipbuilders and ship owners, although it antagonized other countries.† British sea power played a key role in defeating Napoleon Bonaparte. After the Napoleonic Wars ended in 1815, Great Britain was the dominant world power until 1914.

England's participation in two world wars ended its dominance, although England is still among the world's 10 wealthiest nations. The Second World War helped make the United States the world's strongest country, again because of manufacturing. As the "Arsenal of Democracy," the United States had to build new factories, and run existing ones overtime, to make weapons and equipment. The other belligerents were doing the same, but the United States was the only nation out of reach of enemy bombers. England and Germany did their best to destroy each other's factories, while the United States devastated Japan's. At the end of the war, the United States was the only country with a strong industrial base. During the 1950s and 1960s, an American family could often afford a home and two cars on one worker's income.

The party ended in the 1970s after Germany and Japan rebuilt their industries. The Germans and Japanese had to replace the equipment they lost in the war. Once they did this, they had new factories with which to challenge older ones in the United States. In the past couple of decades, American workers have seen their real incomes decline. Again, we see a clear lesson: National power, whether economic or military, comes from manufacturing capability.

---

*"The Act of Navigation, therefore, very properly endeavors to give the sailors and shipping of Great Britain the monopoly of the trade of their own country in some cases by absolute prohibitions and in others by heavy burdens upon the shipping of foreign countries. The following are the principal dispositions of this Act.

"First, all ships, of which the owners and three-fourths of the mariners are not British subjects, are prohibited, upon pain of forfeiting ship and cargo, from trading to the British settlements and plantations, or from being employed in the coasting trade of Great Britain."

—Adam Smith (1723–1790), *An Inquiry into the Nature and Causes of the Wealth of Nations.* http://www.duke.edu/~atm2/index.html, Web page provided by Aaron Miller, a student at Duke University (in 1996).

†"Since the publication of this pamphlet in England, the commerce of the United States to the West Indies, in American vessels, has been prohibited; and all intercourse, except in British bottoms, the property of and navigated by British subjects, cut off . . .

"America is now sovereign and independent, and ought to conduct her affairs in a regular style of character. She has the same right to say that no British vessel shall enter ports, or that no British manufactures shall be imported, but in American bottoms, the property of, and navigated by American subjects, as Britain has to say the same thing respecting the West Indies. Or she may lay a duty of ten, fifteen, or twenty shillings per ton (exclusive of other duties) on every British vessel coming from any port of the West Indies, where she is not admitted to trade, the said tonnage to continue as long on her side as the prohibition continues on the other."

—Thomas Paine, "A Supernumerary Crisis," *Common Sense*, New York, December 9, 1783. Converted to HTML by Danny Barnhoorn for The American Revolution, an HTML project. http://ukanaix.cc.ukans.edu:80/carrie/docs/usdocs.txt/crisis13b.html

The Organization of Petroleum Exporting Countries (OPEC) is in the same position as Spain and Portugal were in the fifteenth century. OPEC members have a valuable natural resource to export, but little manufacturing capability. Despite its wealth, Saudi Arabia depends on the United States for weapons like fighter aircraft. The Saudis even depended on the United States and Great Britain for protection during the Gulf War. Iraq, another oil exporter, had to buy its weapons from the Soviet Union. Meanwhile, the Israelis, who have little oil, had to look to manufacturing for their livelihoods. Israel manufactures its own tanks and aircraft, as do France, England, and Germany. A country that lacks manufacturing capability must rely on outsiders for security.

Here is another way to compare manufacturing capability and natural resources. Suppose that "we" own factories that turn oil into plastics and chemicals, and "they" own an oil well. We can buy their oil, turn it into chemicals and plastics, and sell the goods for a profit. We then use some profit to buy more oil and repeat the process. Their source of wealth—oil— eventually runs out. Ours—the ability to create wealth—does not. (In practice, we may have to repair or replace aging equipment.)

Asia is poor in natural resources. Only about 10 percent of Japan's land is suitable for agriculture, and Korea is mountainous and resource-poor. Since the nineteenth century, Japan has rapidly expanded its manufacturing capabilities. Japanese imperialism stemmed partially from a quest for raw materials to feed Japan's factories and markets for Japanese goods. South Korea is looking to manufacturing to build national wealth and is exporting automobiles and construction equipment. Korean cars are not yet comparable to American, Japanese, or German cars. The Hyundai is an inexpensive economy car, like the Toyota was 25 years ago. Where will the Hyundai be in the next five or so years?

Mainland China is rapidly expanding its manufacturing capability, and we are seeing more Chinese goods in stores. Low labor costs give Chinese factories a cost advantage over American, Japanese, and European ones.

The Importance of Quality

Remember, quality helps us sell our products and stay in business. Quality can help us take away the competitor's business.

If mainland Chinese workers earn a dollar an hour, what prevents them from displacing American workers? Technology is one factor. Mainland China cannot yet produce high-technology goods that will sell in the United States, Japan, or Europe. China cannot make a camera that will compete with

Japanese cameras, nor can it make a Pentium computer chip. China cannot even make a competitive automobile. Communism was the problem, since it doesn't encourage creative and competitive manufacturing. East Germany had the same problem with its Tribant automobile. West Germans said its two tailpipes allowed the owner to use it as a wheelbarrow. When the Berlin Wall came down, Tribant standard features included "five East Germans and a map of West Germany." No one wants a Russian automobile either.

Russia is no longer communist, and its educational system has always been good. China is abandoning communism too. These countries will want a piece of the world's economic action very soon, and they'll get it.

How can advanced nations with high labor costs compete against cheap labor? Technology is one competitive advantage, but not all products are high technology. For example, most of our clothing comes from foreign countries like India and Thailand. Quality, however, is a competitive advantage that applies to *every* product.

- When prices are equal, good quality beats poor quality.

  —Even when prices aren't equal, good quality often beats poor quality.

- The price tag doesn't account for the product's lifetime cost to the customer. Less expensive, but poorer quality goods can be more expensive in the end.

  —The customer may have to replace them more often.

  —The customer may have to repair them more often.

  —When they stop working, poor quality goods can cost the customer money. This is especially true if the customer uses them to make goods or provide services: The industrial customer's machine stops making widgets, or the airline's airplane can't fly.

- Poor quality raw materials or parts can make the job harder or even ruin the final product.

  —Consider the results of having a 1 cent fastener (screw, bolt, and so on) fail in a $1000 system.

  —Remember the story "For Want of a Nail."

- Inferior raw materials or parts might not perform their functions well.

- Good quality during the manufacturing processes usually lowers the manufacturing costs. This means that high quality goods can carry lower price tags than poor quality goods.

What Is Quality?

Quality is a vital instrument for capturing and holding market share. Quality means *fitness for use* as defined by the customer.

Some people view advertising as slick and glamorous and manufacturing as dull and uninteresting. Quality and customer satisfaction, however, come from frontline workers, not the advertising office. Quality and customer satisfaction are critical factors in capturing and keeping market share. Let's compare them to advertising and promotion.

Advertising may help us get new customers. Quality helps satisfy existing customers and keep their loyalty. *It costs five to seven times as much to get a new customer as it does to satisfy and keep one* (Struebing 1996). Therefore, spending $1000 on quality is like spending $5000 to $7000 on advertising and promotion. The story doesn't end here.

Happy customers recommend our product or service to their friends, and that's free advertising. Struebing cites Sheila Kessler (1995), who writes that a lifetime customer of Minute Maid is worth $1 million. This assumes that a customer buys Minute Maid products throughout his or her lifetime and recommends them to others. Under this assumption, a lifetime customer of Delta Airlines is worth $1.5 billion. Now let's examine unhappy customers and their reactions to poor quality.

> I command that the owner of the Tula factory Kornila Beloglazov be flogged and banished to hard labor in the monasteries. He, the scoundrel, dared to sell to the Realm's army defective handguns and muskets.
>
> Foreman alderman, Frol Fuks should be flogged and banished to Azov, this will teach him otherwise than to put trademarks on faulty muskets.
>
> If a stoppage occurs among the troops during combat* due to oversight by the secretaries and scriveners, the latter should be flogged on their naked parts without mercy. The master gets 25 whips and a fine of 10 rubles per one faulty gun. The foreman should be flogged until he loses consciousness. The elder secretary should be enlisted as a warrant officer.

---

*Under the cost of quality model, this is an *external failure* or *field failure*. Chapter 2 will examine cost of quality.

> The secretary of the rank should be stripped to become a copier. The scrivener should be denied his Sunday glass of vodka for one year.
>
> —Decree of Tsar Peter I, 11 January, 1723
> (Juran 1995, 390–391)

Angry customers cannot flog suppliers and sellers today, but they can impose other consequences. These are far more costly than warranty service and product returns imply. Angry customers tell their friends, "Brand X sells low-quality, unreliable junk, and their service is trash." If we're Brand X, we can pour millions of dollars into slick television or magazine ads, and it won't do any good. If people see 10 or 20 of our ads, but their neighbor says "I tried Brand X, and it's junk," they won't buy from us.

It doesn't have to be a neighbor, either. Publications like *Consumer Reports* test and rate products impartially and receive feedback from consumers. One of the authors has seen many television and magazine ads for a luxury automobile. *Consumer Reports* gave it low ratings for reliability (repair frequency). The author wouldn't consider buying it even if he wanted to spend that much money. Therefore, the manufacturer is wasting its money on the advertisements. Meanwhile, a high rating from *Consumer Reports* can help sell the product.

Juran and Gryna (1988, 2.8) define quality as "fitness for use." A quality product or service meets the customer's needs. The customer expresses some needs through specifications, which are often numerical or quantitative. There are, however, other factors in fitness for use (Juran and Gryna 1988, 3.7).

- Operating costs, such as an automobile's fuel economy or the energy efficiency ratings for refrigerators and air conditioners

- Maintenance costs, such as tune-ups, oil changes, and wheel alignments

- Downtime
  - This can be very costly when the customer relies on a piece of manufacturing equipment or service equipment like a passenger airplane. If it isn't working, the customer isn't earning money.
  - In consumer products, it's annoying and frustrating.
  - Maytag uses this as a selling point: the repairperson has nothing to do.

- Factors that include safety and ease of use

In manufacturing, our first job is to meet the customer's specifications. The product design department will (ideally) design the other features into the product. A good design will make the product reliable, easy to use, and cheap to operate.

Manufacturing personnel can and should work with designers and product developers to improve quality. *Design for manufacture* (DFM) is one such cooperative activity. DFM means designing the product so that it is easy to manufacture. This helps improve quality and reduce production costs, thus making the products more competitive.

What Are Specifications?

Manufacturing's primary job is to meet the customer's specifications. Note that customers may be internal or external.

Specifications and standards resulted mostly from the Industrial Revolution in the late eighteenth and early nineteenth centuries. Before the nineteenth century, there were few specifications for any product. Most businesses were trades like blacksmithing, carpentry, shoemaking (cobbling), barrel making, glass blowing, weaving, and so on. Trade workers made and sold the products and relied on their reputations to assure sales. They usually applied a personal trademark to the item, which was an incentive to do good work. In those days, few people traveled far from their homes. Many people lived their entire lives within 10 or 20 miles of their birthplaces. Therefore, most of a trade's business came from fellow townspeople. This made a reputation for good workmanship especially important.

The customers judged quality themselves. Did the plowshare break or give good service? Were barrels tight enough to contain liquids securely? Did shoes fit well, and did they last a long time?

There were legal or official standards for weights, because people sold products by weight. There was some standardization for military weapons, since they had to accept the same ammunition. Even these standards, however, involved weight. A "six-pounder" cannon fired a six-pound cannon ball. Since a six-pound iron sphere has a specific diameter, this effectively defined the cannon's size. The *gauge* of a shotgun—a term that survives today—is the number of lead balls per pound. For example, twelve 12-gauge balls weigh one pound. Only in the nineteenth century did manufacturers begin specifying gun sizes by diameter (inches or caliber).

Specifications and standards allow mass production. The development of interchangeable parts in the early nineteenth century required standards and specifications. Before interchangeable parts, skilled trade workers custom-built each item. For example, a part from one machine would not always fit another of the same kind. If a machine needed a replacement part, a blacksmith had to make it or file one to fit.

Interchangeable parts made mass production possible. A factory could order 1000 nuts and 1000 bolts, with confidence that any bolt would fit any nut. This required the parts to meet specifications. These specifications included length, diameter, and thread geometries. This led to a demand for instruments that could measure dimensions accurately.

Specifications and standards promoted interstate and international trade. Specifications and standards arrived in time to help railroads promote trade. Before the railroad, interstate travel was difficult. Stagecoaches and Conestoga wagons were subject to highway robbery. A broken wheel or lame horse was more than an inconvenience; it was a life-threatening emergency. The railroad made travel routine and allowed interstate commerce. How could a customer be confident in a product if he or she did not know the craft worker who made it? Standards and specifications promoted this confidence, allow with interstate and international trade.

Specifications often include the following:

- Dimensions
    - —Height
    - —Width
    - —Length
- Purity
    - —"No more than 10 ppm (parts per million) impurities"
    - —Ivory Soap's selling point: "99.44% pure"
    - —*Extremely* important in many chemicals for the semiconductor industry
- Viscosity
    - —Paints and coatings
    - —Photoresists for the semiconductor industry
- Weight (for example, for products sold by weight)
    - —Savvy cowboy or backwoodsman to "city slicker" merchant: "Git yer thumb off that scale afore I blow it off!"
    - —"Watering stock" means making cattle or pigs drink a lot of water before selling them to a meat packing factory.
    - —When you buy meat, how much is bone? What about fish sold by weight with the heads still attached?

—Scales must meet certain standards. A scale that is "not legal for trade" means that merchants cannot use it to weigh items.

- Electrical properties

—Resistivity of silicon wafers for semiconductor manufacturing

Figure 1.1 shows how specification limits are like a target. Shots that are in the target are good, and products that meet specifications are good. Shots that miss the target, and products that don't meet specifications, are no good.

The Japanese quality expert Genichi Taguchi says that the goalpost, or good/bad description, is inadequate. He emphasizes that it is best to hit the center of the target. A product whose measurement is exactly between the specification limits is better than one near a limit. Taguchi defines the *loss function* as $L(x) = k(x - m)^2$, where x is the measurement, k is a constant, and m is the bull's-eye or nominal. Alternately, the cost of missing the bull's-eye is proportional to the square of the distance. Similarly, in target shooting and archery, a bull's-eye is worth more points than a hit near the target edge. Figures 1.2 and 1.3 compare the traditional model to the Taguchi model. Figure 1.2 assumes a perfect unit is worth 10 points (or dollars, and so on). Figure 1.3 compares the models graphically.

In most commercial activities, however, the good/bad or goalpost model applies. The customer will pay full price for any part that meets specifications and nothing for one that doesn't. Also, the purpose of statistical process control (SPC) is to get as close as possible to the bull's-eye. Therefore, it doesn't matter whether we choose the traditional specification model or the Taguchi model. We want to get as close to the bull's-eye (nominal dimension) as possible.

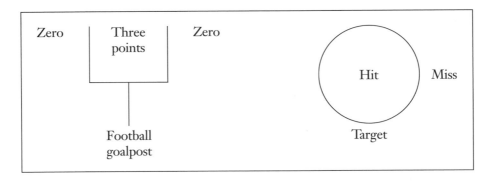

Figure 1.1. Specification limits.

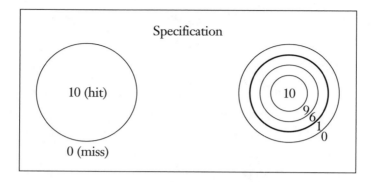

Figure 1.2. Taguchi model versus traditional model (targets).

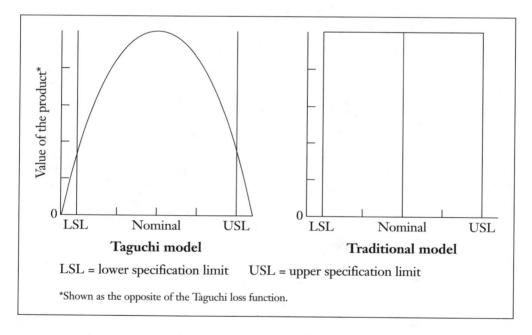

Figure 1.3. Taguchi model versus traditional model.

Who Are Customers?

We usually think of customers as the people or organizations that buy our company's products. These are *external customers;* however, we also have *internal customers.* These are the people or departments in our own company to whom we supply a product or service. Internal suppliers are people or departments in our own company that provide us with products or services. *All work in process (WIP) passes from an internal supplier to an internal customer.*

Like external customers, internal customers have specifications and needs. Meeting their requirements helps them meet those of internal customers further down the line. In turn, this helps the company meet the needs of external customers.

The Manufacturing Worker's Role in Quality

Today's manufacturing workers play a critical role in assuring and improving quality. Modern management experts agree that the frontline manufacturing workers' skills and experience are vital to an organization's competitiveness. Management consultant Tom Peters, the coauthor of *In Search of Excellence* (Peters and Waterman 1982), is one of today's foremost experts on business management. Peters says that frontline workers know more about their job than anyone else. An engineer or scientist may know more about the technology behind the job, but he or she does not do the job eight hours a day. A manager may know more about how the job fits into the organization's mission, but the manager doesn't do it eight hours a day either. The people who have their hands on the materials and equipment learn intimate details about the task.

Frontline workers often develop informal techniques that can improve quality or productivity. Frank Gryna defines a *knack* as "a small difference in method which accounts for a large difference in results" (Juran and Gryna 1988, 22.57). Knacks come only from hands-on experience with the job. (Gryna is coeditor of *Juran's Quality Control Handbook*, which is among today's foremost references on quality.)

Workers, however, should not deviate from approved practices, because changing the method can damage the product. Instead, workers who think they can improve on the procedures should seek approval through the proper channels. This often means involving the process engineers, technicians, and other workers. The engineers or technicians may help by designing a controlled experiment to test the new technique. The experiment can show whether the change improves quality or productivity or causes harm. Often, however, the engineers may simply approve the change because there is no potential for harm. The change then becomes part of the official procedure. Harris Semiconductor calls this procedure an *operating instruction* or OI. Other companies may call it a process specification, process instruction, work instruction, or similar name. When the change becomes part of the official procedure, everyone is aware of it and can benefit from it.

Self-Directed Work Teams

Self-directed work teams (SDWTs) plan and carry out their own work assignments. They also initiate and carry out projects to improve productivity and quality. The star organization (Figure 1.4) for a SDWT helps assure performance of critical tasks and avoids saddling the team leader with all the work.

Harris Semiconductor's SDWTs have no foremen or supervisors. At the Mountaintop plant, six or more SDWTs of 6–12 employees report to a manufacturing leader. The manufacturing leaders report to the manufacturing manager, who reports to the plant manager.

Teams start and carry out projects to improve productivity and quality. At Mountaintop, each team has a bulletin board that shows team projects, training, and other team activities. Each team selects its own leader, who coordinates team activities like quality improvement projects. Unlike a foreman, the leader does not have formal authority. He or she is like the leader of a small professional, civic, or volunteer organization. The leader usually contacts engineers and other technical support people for help. The leadership position often rotates among team members.

Initiatives by frontline workers at Mountaintop have improved Harris' profitability and productivity. These projects have reduced scrap and rework and improved product yields. One initiative changed a rinse process to reduce stains on semiconductor wafers, and this increased the product yield.*

---

*In semiconductor manufacturing, this is the percent of the wafer that is good. Each wafer has hundreds or even thousands of rectangular die, chips, or pellets. (These words are synonyms.) Each die is a separate electrical device, like a transistor or rectifier (diode). Processing occurs while the die are in wafer form. At the end of the process, a diamond-edged saw separates the die for packaging and shipment. Computer companies like Intel and IBM use similar processes to make computer chips.

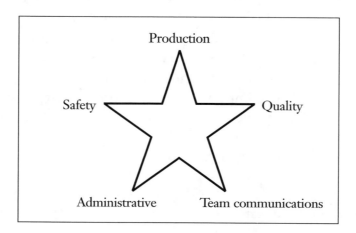

Figure 1.4. "Star" organization for a self-directed work team.

Another initiative equipped automatic wafer coaters with photoresist* supply sensors. This allowed the equipment to use all of the costly material in each bottle (Murphy and Levinson 1996).

SDWTs have been successful at other Harris locations, including the large plant in Palm Bay, Florida. Their successful introduction depended on skillful management of organizational change. That is, a company cannot merely say, "We now have self-directed work teams." It must put the organizational and social structures in place to support the change, and it must build trust between all organizational participants (Rose, Odom, and Pankey 1996).

The Mountaintop plant recently introduced the star organization for SDWTs. The star organization is like a professional, civic, or fraternal organization. This team structure helps assure performance of critical tasks, although the organization has no formal authority structure. It avoids saddling the team leader with all the administrative tasks. Although no one can tell anyone to do anything, the informal authority of the situation promotes mutual acceptance of responsibilities.

How can a system without an authority figure, like a foreman or supervisor, be effective? Juran and Gryna (1988, 22.60) cite the "law of the situation" as follows. "One *person* should not give orders to another *person*. Both should take their orders from the 'law of the situation.'" The task, or situation, is the authority figure that tells everyone what to do.

A typical professional, civic, or fraternal organization usually has a president (team leader), secretary, treasurer, newsletter editor, program director, and so on. This arrangement associates a specific person with each critical task. The star organization does the same in a SDWT. This avoids the problem of, "if everyone is responsible, no one is responsible." The industrial folk tale about Everybody, Somebody, Anybody, and Nobody explains this problem. In the story, a group had to do a simple task. It was Everybody's job, Anybody could have done it, Everybody thought that Somebody would do it, but Nobody did it. The star arrangement prevents the story from ending this way. Typical star assignments appear in Figure 1.4; however, a star can have any number of points, and it can include other activities.

---

*A photoresist is a photosensitive film that goes on the wafer. A process similar to photography produces microscopic electronic devices on each die. The photoresist, like most chemicals for the semiconductor industry, is expensive. This is because these materials must be especially pure.

Customer Contact Teams

Customer contact teams (CCTs) include manufacturing workers, engineers, and managers. Most of the members are manufacturing workers. CCTs meet directly with manufacturing workers on the customer's shop floor. This approach takes advantage of the customers' workers' intimate knowledge of their manufacturing process. It also shortens the customer-supplier communication path. Figures 1.5 and 1.6 compare the traditional customer-supplier communication path to the CCT communication path.

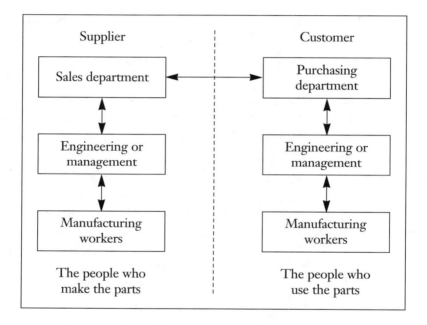

Figure 1.5. Traditional customer-supplier communications.

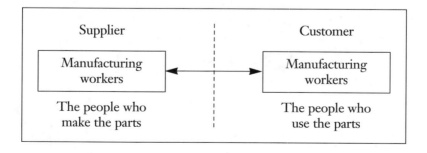

Figure 1.6. CCT communication path.

W. Edwards Deming (1900–1993), one of the twentieth century's foremost quality experts, said to "break down [organizational] barriers between departments." This means that people in different parts of the company must be free to talk and work with each other. Douglas MacArthur invited Deming to postwar Japan to help the Japanese rebuild their industry. Deming taught the Japanese SPC and other techniques for improving productivity and quality. He was largely responsible for Japan becoming one of the three leading economic powers (with the United States and Germany) today. The Japanese named an industrial quality award, the Deming Prize, after him.

The CCT extends Deming's advice by removing barriers between the customer and the supplier. Peters refers to a barrier-free organization as "porous." Porosity makes organizations flexible and responsive to diverse customer needs. Peters (1988) specifically says that frontline workers should interact with customers and suppliers. "The 'average' person . . . will routinely be out and about—that is, firstline people communicating directly with suppliers, customers, etc. Who is the person who best knows what's wrong with defective suppliers? Obviously, the frontline person who lives with the defective item eight hours a day" (p. 14). Peters also says that frontline workers can work with suppliers to improve quality or productivity (p. 18).

Recall that specifications do not always reflect all the customer's needs. Workers at a major customer's plant were unhappy with Harris parts, although the parts met specifications. Manufacturing workers met with the customer's workers and identified the problem. Some parts had cosmetic (visual appearance) defects that the specifications did not cover. These defects interfered with the customer's machine vision systems. The CCT solved the problem, and the customer was very happy with the results.

The CCT went on to improve a Harris product for this particular customer. The customer buys electronic devices in the form of die. Die, chips, or pellets are tiny rectangles of silicon that act as transistors or resistors. The manufacturer produces them on silicon wafers, one of which can hold hundreds or thousands of die. When the manufacturing process is complete, a diamond saw cuts the wafer into individual die. First, however, an electrical tester evaluates each die individually and marks it with an ink dot if it is defective.

Harris normally discards the bad die and places the good ones on tape reels for shipment. This customer, however, uses die on wafers instead of reels. The customer's die bonders transfer the die from the wafers to copper heat sinks. The die bonders detect the ink marks on the bad die and do not use them. Skipping the bad die, however, takes time and reduces production speed. The CCT devised a process to take good die and put them on Mylar tape in the shape of a wafer. The customer's die bonder can accept the die in this form, and, since all of them are good, the process runs at full speed. Figure 1.7 shows the process improvement.

Since then, CCTs from the Mountaintop plant have successfully worked with two other large customers. Feedback from one contact led Harris to increase the tape reel size by 67 percent. This increased the customer's time between reel changes and reduced the customer's equipment setups. This improved the customer's productivity.

Quality Improvement Teams

The mission of a quality improvement team (QIT) is to solve a problem or make an improvement. A *quality circle* is another name for a QIT. Its key feature is its involvement of people from several parts of the company. A QIT

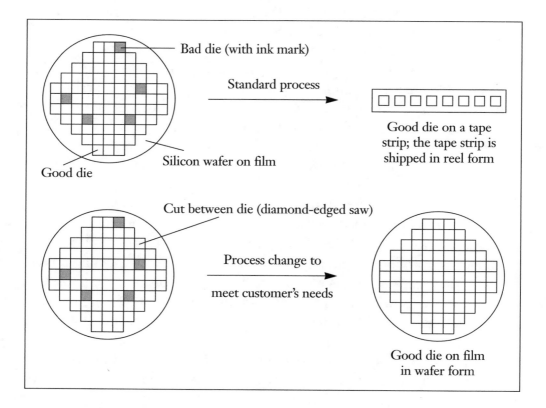

Figure 1.7. CCT's process change.

crosses organizational boundaries, and a CCT is actually a QIT that works with a customer. QITs often involve the following:

- Manufacturing workers

- Manufacturing engineers and technicians

- Product designers

  —A change in the product's design may make it easier to manufacture. This is part of DFM.

- Industrial statisticians

  —Design of experiments is a method of scientifically testing process changes to see if they improve quality.

CHAPTER TWO

Tools for Improving Productivity and Quality

Friction

Friction refers to seemingly minor annoyances that add up to major problems. We often don't notice them because we can work around them. They eventually become an accepted part of the job. These minor annoyances and inefficiencies can seriously undermine quality and productivity. Frontline workers are in the best position to discover and correct sources of friction.

You have probably read about friction and seen examples in everyday life. We'll start with Benjamin Franklin's famous story "For Want of a Nail." "For want of a nail the shoe was lost; for want of a shoe the horse was lost; and for want of a horse the rider was lost." The absence or failure of an inexpensive and seemingly unimportant part—a simple nail—has serious consequences. Franklin's story has several modern counterparts. The following examples include boring, mundane, and seemingly unimportant items like nails, nuts and bolts, screws, gaskets, and fan belts.

Some versions of Franklin's story happen during a battle. The rider is a high-ranking officer or a messenger with a vital message. This causes the army to lose the battle, which allows the enemy to overthrow the kingdom.

There have been problems with counterfeit fasteners,* such as bolts and capscrews. For example, unscrupulous vendors sold SAE grade 7 fasteners as grade 8, which is the highest grade.

- The grade 7 bolt's minimum tensile strength is 88.7 percent that of the grade 8 bolt (Baumeister, Avallone, and Baumeister 1978, 8-28, Table 30, "Physical Requirements for Threaded Fasteners").

- Fasteners hold airplanes and automobiles together.

- "For want of a fastener, the airplane's wing was lost . . ."

---

*"FASTENER QUALITY ACT: Public Law 101-592, Overview and Update, 7/14/95"

"During 1985 and 1986, reports of substandard, mismarked and/or counterfeit fasteners triggered a nationwide investigation that ultimately involved five countries and resulted in Congressional hearings and the introduction of HR-3000, the Fastener Quality Act, in the summer of 1990." (National Institute of Standards and Technology, Web page at http://ts.nist.gov/ts/htdocs/200.html)

Reputable manufacturers have published advertisements warning chemical companies about counterfeit gaskets and seals. For example, unscrupulous suppliers may put reputable brand names on low-quality seals. These suppliers are like the counterfeiters who mark "Rolex" on cheap watches. Gaskets and seals are what keep corrosive, flammable, and toxic chemicals inside pipes and valves.

Several years ago, one of the authors bought a computer disk drive. It came without the screw for attaching it to the computer frame. The author had to visit two hardware stores to find the right screw. The supplier's representative saw nothing wrong with this. He said, "We assume that the buyer knows how to install the disk drive." How will this affect the author's future consideration of this supplier? "For want of a screw, the customer's goodwill was lost."

In 1987, one of the authors bought a new car. Shortly afterward, the power steering pulley came off and flew into the fan. It took a couple of other belts with it, including the water pump belt. A car can be driven without power steering, but not without a water pump.

In each of these examples, failure of a very mundane and seemingly unimportant item has serious results. Here is another familiar example. A supermarket fails to program a product's bar code into the price scanner. The cashier cannot ring up the item and must call someone to check the price. The checkout line stops while this happens. This is very annoying to the customers, and it makes extra work for the supermarket's personnel. Some customers have very little patience for this and often tell the cashier to put the item aside instead of waiting for the price check. The supermarket loses the sale.

Here are some more examples from the semiconductor industry. A bad semiconductor package (transistor cap or stem) jams the packaging machine. This causes downtime while the operator clears the jam. The cap or stem is a simple metal piece that costs about a dime. The automatic packaging machine costs around half a million dollars.

Workers are trying to do a job and can't find the screwdriver they need. We will discuss the 5S-CANDO program, which prevents this type of incident. 5S-CANDO is a systematic activity for organizing and arranging the workplace.

A brief (1–2 second) power outage stops the computer system and ruins half an hour's work. An uninterruptible power supply (UPS) prevents this type of incident by giving users a few minutes to save the work.

Equipment for handling thin, brittle, and expensive silicon wafers causes breakage. Tweezers were acceptable for handling small (2″ or 3″) wafers, but not larger wafers. Today, semiconductor manufacturers use 6″ and 8″ wafers. These wafers' greater weight requires more tweezer pressure. Too much pressure will break the wafer. If there is too little pressure, the wafer will slip out of the tweezers, land on the floor, and break. Vacuum pencils have replaced tweezers in most semiconductor factories.

Electrostatic discharge (ESD) can easily ruin a finished transistor. ESD is a fancy name for what happens when we walk on a thick carpet and then touch a metal object. It's not something we pay much attention to, since we feel only a brief minor shock. During the 1970s, one of the authors lost a hand calculator to ESD. (Newer calculator designs may protect their circuits from ESD.)

A bad electrical probe card causes the electrical tester to reject an entire wafer. The wafer must be retested. Figure 2.1 shows that each wafer has several hundred die, chips, or pellets. (These three words have the same meaning. For simplicity, the die in the figure are larger than real ones.) Each die is an electrical device like a transistor, diode, or rectifier. At the end of the process, a saw cuts the wafer into individual die. Each die goes into a package, the most familiar of which is a three-pin transistor package. The electrical tester checks each die while it is still on the wafer. It marks bad ones for rejection by the automatic packaging equipment.

It would be convenient to have a single word to describe all of these incidents. Carl von Clausewitz (1976), a Prussian general, provided one in his famous book *On War*.

> Friction, as we choose to call it, is the force that makes the apparently easy so difficult . . . Countless minor incidents—

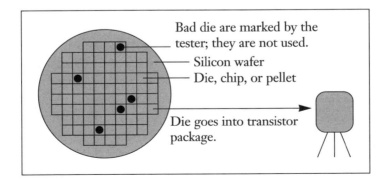

Figure 2.1. Semiconductor wafer and final product.

the kind you can never really foresee—combine to lower the general level of performance, so that one always falls far short of the intended goal. (Book 1)

Peters (1987, 323) echoes this principle in *Thriving on Chaos.* "The accumulation of little items, each too 'trivial' to trouble the boss with, is a prime cause of miss-the-market delays." Friction is the source of Murphy's Law: "Anything that can go wrong, will." Frontline workers who do a job eight or more hours a day are in the best position to discover sources of friction.

Friction is insidious because we can often work around it. This means fixing breakdowns, reworking parts, sorting (rectifying, detailing) good parts from bad ones, and so on. The friction eventually becomes a routine part of the daily work. People stop noticing it, but it doesn't go away. It continues to reduce the operation's efficiency and effectiveness and continues to make the job frustrating and less enjoyable.

Nakajima (1989, 43) describes chronic and sporadic problems. The latter usually get attention, while people ignore the former. A chronic problem recurs frequently, but each incident causes little harm. It is easy to work around the problem. Therefore, operators rarely tell supervisors or engineers about it. No one deals with its causes, and it continues to make trouble.

Here is a rule for identifying friction. If it's frustrating, a chronic annoyance, or a chronic inefficiency, it's friction. We will now consider some programs and activities that suppress friction and its effects.

Total Productive Maintenance and Preventive Maintenance

Total productive maintenance (TPM) is a program for assuring continuity of operations and for reducing defects. Continuity of manufacturing operations means the equipment is available when we need it. We are familiar with continuity of operations in everyday life. Does the automobile start when you turn the key? Does it keep running? Can you rely on appliances like the washing machine and dryer? How reliable is the power company? How reliable is your telephone service?

Equipment breakdowns are very serious in manufacturing. This is particularly true in multistep processes, where failure of one workstation can shut down the following operations. They are also serious in service industries like banking and travel. "The computer is down" is a common reason (or excuse) for delays in bank transactions and travel reservations. Failure of an air traffic control computer can cause long delays at airports. Depending on the application, even 99 percent availability may not be good enough.

Juran and Gryna (1988, 3.20–3.22) discuss the life cycle cost of durable goods. Durable goods include manufacturing equipment, vehicles, and appliances, while consumable goods include food, fuel, materials, and services. Cost accounting systems treat consumable items as expenses, while they depreciate durable items. A consumable item's cost is its purchase price. The true cost of a durable item includes far more than its purchase price. The life cycle cost includes the following:

- Purchase price
- Operating costs (fuel economy, power efficiency)
- Repair and maintenance costs
- Insurance costs
- Disposal costs

A 1996 advertisement by Kitz Corporation uses life cycle costs as a selling point. The ad shows pictures of valves with red labels such as "Corroded," "Cracked," "Leakage," "Who bought this?" and "Imitation." The caption says, "So much for red tag sales." The ad warns, "Too often, poor quality hides behind a low price." A chemical manufacturer can get a low price for a cheap valve, but what happens if it malfunctions? The costs could include repair or even a plant shutdown. The cheap valve is a poor bargain when you consider its total cost of ownership.

Tantalum is a corrosion-resistant, but costly metal. A tantalum heat exchanger for a chemical process may be three times as expensive as a graphite one. A buyer's immediate reaction might be, "Buy the graphite one because it's much cheaper." The tantalum unit, however, requires little maintenance and may need replacement every 15 years. The graphite unit requires a lot of maintenance and a new one may be needed every five years. Tantalum doesn't crack, but graphite does, and this causes downtime (Burns and Kumar 1996). We will see that downtime is more expensive than the cost accounting system shows.

Life cycle costs reflect money that we have to pay to own and use the item. They do not, however, reflect *opportunity costs*. These costs refer to losing opportunities to use the durable item to make money. Consider a passenger airplane that misses a flight because of mechanical problems. Suppose it costs $2000 to repair the airplane, and the airline would have earned a $5000 profit from the flight. The costs of repairing the airplane show up in the airline's cost accounting system. The lost opportunity to earn a profit by transporting passengers doesn't show up. Finally, if the cancellation inconveniences the passengers, they may choose a different airline the next time. Similarly, in a manufacturing operation, the actual cost of fixing equipment may be the least of our problems. This is especially true if the breakdown keeps us from making products to sell to customers. We will miss a profit, and we may inconvenience customers who are relying on our delivery schedule. In summary, breakdowns of manufacturing or service equipment can be far more costly than the repair costs suggest.

For example, suppose we have to choose between the following machines. Machine B is 50 percent more expensive than Machine A. The machine is for an operation that limits the productivity of the manufacturing line (constraint). When the operation is working, it earns the company $500/day. Which machine is better?

	Machine A	Machine B
Annual depreciation*	$20,000	$30,000
Annual electrical cost	$3000	$3000
Annual repair cost	$4000	$1000
Daily earnings	$500	$500
Downtime, days/year	22	2

The life cycle cost includes the operating cost (electricity) and the repair costs. Every day that a machine is unavailable, the company loses the opportunity to make $500. We can't easily quantify the effect of disappointing customers who wanted the product. We know they don't like it. If they buy from a competitor because we can't fill their order, it gives the competitor a chance to get our business. Suppose a customer normally shops at Store A and has never been to Store B. Store A sells out of something that the customer wants, so he or she has to visit Store B. This customer may like Store B better. Alternately, consistent customer satisfaction can prevent the customer from even looking in the competitor's store.

Call the dollar value of customer goodwill x (per day). Here are the total annual costs for the machines. Although Machine B costs more to buy, it costs much less to own.

---

*The purchase price, distributed over the machine's lifetime.

	Machine A	**Machine B**
Purchase cost (annualized)	$20,000	$30,000
Life cycle costs, excluding the purchase cost	$7000	$4000
Opportunity costs	$11,000	$1000
Loss of customer goodwill	$22x$	$2x$
Total	$38,000 + 22x$	$35,000 + 2x$

A workstation's *availability* (Juran and Gryna 1988, 2.9, 13.31–13.42) is:

$$\text{Availability} = \frac{\text{Uptime}}{\text{Uptime} + \text{Downtime}} \times 100 \text{ percent}$$

Uptime	**Downtime**
• The unit is operating	• Under repair
• The unit is idle, but available for use	• Waiting for parts
	• Waiting for paperwork

Overall equipment effectiveness (OEE) measures the net effectiveness of a manufacturing tool. Table 2.1 shows its components.

OEE = Availability × Operating efficiency × Rate efficiency × Rate of quality

Table 2.1. Components of overall equipment effectiveness.

The *availability* is the equipment's percentage of uptime. When the equipment is up, it is available for use.	$\dfrac{\text{Uptime} = \text{Operating time} + \text{Idle time}}{\text{Uptime} + \text{Downtime}} \times 100\%$
The *operating efficiency* reflects the idle time portion of the uptime.	$\dfrac{\text{Operating time}}{\text{Uptime} = \text{Operating time} + \text{Idle time}} \times 100\%$
The *rate efficiency* is the ratio of the unit's actual output to its theoretical output. Losses in rate efficiency result from operation at less than full speed or with partial loads.	$\dfrac{\text{Actual output (pieces/time)}}{\text{Theoretical output (pieces/time)}} \times 100\%$
The *rate of quality* is the percentage of the output that is good. Quality rates of less than 100% result from scrap and rework.	$\dfrac{\text{Good pieces}}{\text{Actual output} = \text{Good and bad pieces}} \times 100\%$

When we multiply these factors, uptime and actual output appear in the numerator and denominator of the result. This allows simplification as follows:

$$\text{OEE} = \frac{\text{Operating time}}{\text{Total time}} \times \frac{\text{Good pieces}}{\text{Theoretical output}} \times 100\% \quad \textbf{(Eq. 2.1)}$$

For example, here are data for a machine for a 24-hour day.

Downtime	2 hours
Idle time	3 hours
Theoretical rate	100 pieces/hour
Actual output	1615 pieces
Rework or scrap	97 pieces

Availability	$\dfrac{(24 - 2 \text{ downtime hours})}{24 \text{ hours}} \times 100\%$	91.7%
Operating efficiency	$\dfrac{(22 \text{ uptime} - 3 \text{ idle}) \text{ hours}}{22 \text{ hours}} \times 100\%$	86.4%
Rate efficiency	$\dfrac{1615 \text{ pieces}}{19 \text{ hours} \times \dfrac{100 \text{ pieces}}{\text{hour}}} \times 100\%$	85.0%
Rate of quality	$\dfrac{(1615 - 97) \text{ good pieces}}{1615 \text{ pieces}} \times 100\%$	94.0%
OEE	$\dfrac{19 \text{ hours}}{24 \text{ hours}} \times \dfrac{1518 \text{ good pieces}}{19 \text{ hours} \times \dfrac{100 \text{ pieces}}{\text{hour}}} \times 100\%$	63.3% 63.3%

Multiplying the percentages in the right-hand column, or using Equation 2.1, produces the same answer. What does the result mean? The machine's effectiveness is only 63 percent of what it would be if it always operated at full speed, without making rework or scrap. Two machines with 95 percent OEEs can do the work of three machines with 63 percent OEEs. If we could get the OEE to 95 percent, we would only have to spend two-thirds as much on equipment.

We will later see, however, that operating and rate efficiencies are critical only for a constraint or bottleneck process. The constraint or bottleneck is the slowest operation in the process, and it limits the overall productivity. Except at the bottleneck, it is acceptable to have idle time and partial loads. Rework and scrap, however, are never desirable. It always costs time and money to replace them.

Preventive maintenance (PM) is a technique for reducing downtime and increasing availability. If the equipment never breaks down, we don't have to worry about waiting for parts or paperwork. We don't have to worry about how long it takes the maintenance worker or machine attendant to fix the machine.

Changing the oil in a car is PM. Since dirt in the oil can gradually damage the engine, changing the oil and filter prevents costly engine repairs. Checking and replacing belts and hoses before they actually fail or break is PM. If the cooling pump belt breaks while on the road, the car is almost undriveable. The engine can be run for a few seconds, turned off, and then the car can coast. This will get the driver to a garage if one is nearby. Otherwise, a tow truck will be needed. It is much better to replace the belts periodically and avoid this situation.

Getting a vaccination is also preventive maintenance. Vaccines for influenza and pneumonia cost from $5 to $20, and it takes a few minutes to get a shot. If people get flu or pneumonia, they may be out of work for several days or even weeks. Evenings and weekends won't be very enjoyable either. Medical costs may easily run into hundreds of dollars. Health insurers that don't cover these vaccines are not very intelligent, even from a self-interest viewpoint.

TPM evolved from PM, and adds improvement-related maintenance and maintenance prevention (MP). TPM includes the five major elements shown in Table 2.2.

Shirose (1992) says that manufacturing workers often overlook stoppages because they are easy to correct. For example, the worker may simply remove a piece that has jammed the machine. This is a perfect example of friction. We overlook the problem because we can work around it. Stoppages, however, point to inherent problems with either the machine or the incoming materials.

The next section describes 5S-CANDO, a program for cleaning and organizing the workplace. 5S-CANDO interacts closely with TPM, since systematic cleaning helps prevent breakdowns and defects. Cleaning also provides an opportunity to inspect the equipment. Finally, a clean workplace makes it easy to detect abnormalities.

Table 2.2. Five major elements of TPM (Shirose 1992, 11).

1. Improvement activities for making equipment more effective	Reduction of 1. Breakdowns 2. Setups and adjustments 3. Idling and minor stoppages 4. Speed losses 5. Defects, rework, and scrap 6. Startup losses (losses due to a setup change or machine adjustment)
2. Autonomous maintenance by manufacturing operators	Manufacturing operators perform routine maintenance activities such as cleaning, inspection, and lubrication. They also learn to recognize and respond to abnormal conditions.
3. Planned maintenance	These activities are similar to the maintenance schedule for an automobile. For example, the owner's manual may recommend an oil change every 3000 miles. There is also a schedule for transmission service, wheel bearing cleaning and repacking, and so on.
4. Improvement of operation and maintenance skills through training	Manufacturing operators learn to perform routine maintenance tasks (autonomous maintenance) and identify abnormalities. Maintenance workers and machine attendants learn advanced skills and techniques. Nakajima (1989, 330) compares the operator to an automobile driver. The driver performs routine tasks such as checking fluid levels and tire pressure. The driver can add oil or transmission fluid or put air in the tires; however, an auto mechanic (the maintenance worker) must perform complex repairs or maintenance tasks. For example, most drivers have neither the training nor the equipment to do a tune-up.
5. Design for maintenance prevention, and early equipment management	This usually happens during equipment design and does not involve manufacturing operators. The idea is to design new equipment to reduce its maintenance needs. Another goal of MP is to make preventive maintenance and repair easy. For example, is it easy or difficult to reach your car's oil dipstick? How hard is it to replace a headlamp or turn signal? Other aspects of MP include design for ease of operation, quality, and safety.

5S-CANDO

5S-CANDO is a set of activities for reducing friction and making the workplace safer. It is a systematic technique for cleaning, organizing, and arranging the work area.

> The 'eathen in 'is blindness bows down to wood an' stone
> 'E don't obey no orders unless they is 'is own
> 'E keeps 'is sidearms awful; 'e leaves 'em all about
> And then comes up the Regiment an' pokes the 'eathen out!
>
> *All along o' dirtiness, all along o' mess*
> *All along o' doing things rather-more-or-less*
> *All along of abby-nay, kul, and hazar-ho*
> *Mind you keep your rifle an' yourself jus' so!*
>
> —Rudyard Kipling, *The 'Eathen*

abby-nay = "not now"   kul = "tomorrow"
hazar-ho = "wait a bit"   Alternate translations: "When
                              I get around to it"

Kipling's poem *The 'Eathen* urged the English recruit to "keep his rifle and himself just so." The British Army often owed its successes to discipline and organization. This included systematic care, cleaning, and maintenance of its equipment. The same activities promote success in manufacturing competition.

Some countries may still resent nineteenth century England's arrogance toward their natives. The English themselves, however, were once the undisciplined "heathens." The ancient Britons wore woad, or blue dye (one size fits all, never needs tailoring or laundering). "Round to it" may have originally meant "sharpening stone." For example, "I'll sharpen my spear when I get around to it." They presumably "kept their sidearms awful" while the Roman Empire's soldiers kept their swords, armor, and themselves "just so." This is why Rome once ruled everything to the south of Hadrian's Wall. The Britons probably learned something from this. Hundreds of years later, someone like Oliver Cromwell or the Duke of Wellington issued every soldier a round to it.

This little item helped the English rule half the world. 5S-CANDO can help a business rule the marketplace. It includes a round to it in the form of systematic, routine activities. 5S refers to five Japanese words for the activities, while CANDO is an English acronym (Table 2.3). Here is a summary of the

Table 2.3. 5S-CANDO.

Japanese 5S	Translation	CANDO
Seiri	Clearing up	Clearing up
Seiton	Organizing	Arranging
Seiso	Cleaning	Neatness
Shitsuke	Discipline	Discipline
Seiketsu	Standardization	
		Ongoing improvement

elements of 5S-CANDO from Harris Semiconductor, Mountaintop's "Can Do Workshop."

Clearing up

1. Remove nonessential items from the work area and develop a system to prevent nonessential items from collecting.

   a. Put red tags on items you believe are nonessential. The purpose is cross-shift communication. Maybe another shift uses the item.

   b. Red-tagged items go in a holding area. If no one claims them after a given time, discard them.

2. Organize items by frequency of use.

   a. Frequent (daily) use; keep at the workstation.

   b. Average (weekly) use; keep near the workstation.

   c. Infrequent (monthly) use; remove from the area.

Arrangement

1. "A place for everything, and everything in its place."

2. Develop an efficient storage system. Get appropriate cabinets, racks, boxes, and so on.

3. Label tools, equipment, and parts, and their storage locations.

4. Make items easy to remove and return.

   a. Example: Hang wrenches on a wall rack. The rack has a place for each wrench. Return the wrench to the correct position after using it.

   b. Example: A socket wrench box may have a specific location for each socket. Returning the socket to the correct position avoids having to look for it the next time anyone needs it.

Neatness

1. Keep the process equipment, tools, walls, and shop floor clean.

2. A major benefit is that many abnormalities and problems reveal themselves.

   a. Example: A pipe has a slow leak. If the leak is over a clean floor, it will be quickly found. If it is over a dirty floor, it may not be noticed.

3. Another benefit is removal of dirt and debris that can interfere with the equipment or damage the product.

   a. In semiconductor manufacturing, even microscopic dust particles can ruin the product.

4. The workplace will be safer and more enjoyable.

Discipline

1. Discipline means standardizing operations and activities and following the standards.

   a. Everyone receives appropriate training.

   b. Employees help create the rules, which gains everyone's support and commitment. Employees help create and modify checklists.

   c. Management shows commitment to CANDO.

   d. Daily checking and cleaning become routine.

Ongoing improvement

1. Generate and adopt new and innovative ideas and aggressive goals.

2. Do not accept abnormalities (friction). Do not let friction become a routine part of the job by "working around it."

   > *Keep away from dirtiness—keep away from mess*
   > *Don't get into doin' things rather-more-or-less!*
   > *Let's ha' done with abby-nay, kul, and hazar-ho;*
   > *Mind you keep your rifle an' yourself jus' so!*
   >
   > —Rudyard Kipling, *The 'Eathen*

Improving Productivity: Synchronous Flow Manufacturing

A factory exists to make money. It can do this only by producing and selling finished goods. Operating efficiencies support this goal only if they improve throughput. A common mistake is to produce inventory solely to maintain operating efficiencies.

Synchronous flow manufacturing (SFM) reduces inventory and improves work flow. It is similar to just-in-time manufacturing.

Does the performance measurement serve the factory, or does the factory serve the performance measurement? Here is an example of the factory serving the performance measurement. A three-step process makes a product. The company uses overall equipment effectiveness (OEE) to measure the work teams' performance.

Step	Capacity, units/hour
1	15
2	12
3	20

Recall that idle time reduces the operating efficiency. Running with partial loads, or at less than full speed, reduces the rate efficiency. To get the top efficiency, there should be no idle time, and the tool should always work at full speed. Suppose that each workstation in this process tries to work at top speed. Figure 2.2 shows what happens.

The first station can make three more parts per hour than the second can process. Running station 1 at full efficiency piles up three units of inventory per hour in front of station 2. Station 3 cannot be more than 60 percent efficient. It can process 20 pieces each hour, but receives only 12 pieces from station 2.

Is the warehouse full of inventory good or bad? Inventory is supposedly an asset, and cost accounting systems treat it as an asset. Suppose that it costs $5 in labor and raw materials for station 1 to process a unit. Here is what it costs to fill the warehouse with a year's worth of inventory.

$$\frac{3 \text{ pieces}}{\text{hour}} \times \frac{\$5}{\text{piece}} \times \frac{24 \text{ hours}}{\text{day}} \times \frac{360 \text{ work days}}{\text{year}} = \$129{,}600$$

This assumes operation on weekends to avoid idle time on the machines!

It costs $129,600 in cash to make the inventory. It is, however, worthless until it is turned into finished goods and sold. We've merely succeeded in tying up $129,600 in inventory. We'd be better off investing the money in the stock market or even in the bank. On average, the stock market gains 12 percent or 13 percent a year. The money we've tied up in inventory is doing nothing, when it could be earning $16,000 or so a year.

There is another danger of tying up cash as inventory. Poor cash flow can drive even a profitable business into bankruptcy. On paper, inventory is a current asset,* and it makes the company's current ratio (current assets/current liabilities) look good. In practice, we can't pay bills with inventory. We

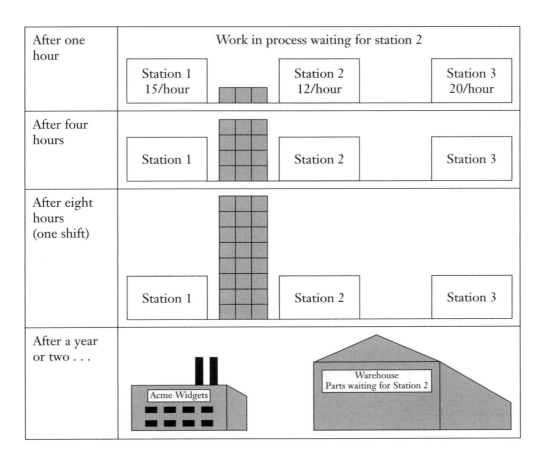

Figure 2.2. Seeking to maximize all equipment efficiencies.

can easily "go pork" (as one author's accounting professor put it) with a warehouse full of current assets.

So far, we can conclude that a multistep manufacturing process cannot work faster than its slowest operation or *constraint*. Trying to make it go faster simply generates piles of inventory in front of the constraint.

SFM is a technique for reducing inventory. Figure 2.3 shows how SFM ties production starts to the constraint. That is, the constraint sets the pace for the entire process. In this example, station 1 would normally make only 12 pieces each hour.

---

*A current asset is something we can, in theory, turn quickly into cash. Current assets include inventory, work in process, accounts receivable, and, of course, negotiable securities and cash. The current ratio is a measurement of a company's economic health.

Liquid or "quick" assets include only cash and negotiable securities. The liquidity ratio, quick ratio, or "acid test" ratio is liquid assets/current liabilities. Therefore, we are better off with cash than with inventory.

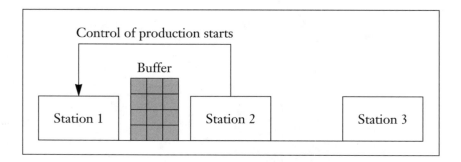

Figure 2.3. Synchronous flow manufacturing.

We must never allow the constraint to stop working. If any other operation loses time, it can make it up by working faster. Time losses at the constraint are irrecoverable.

An equipment breakdown is only one way the constraint can lose time. "Starving the constraint" means idling the constraint by letting it run out of work. The inventory buffer before the constraint assures that this operation will always have work. If the first station goes down, the constraint can work on the inventory. The size of the buffer depends on how long it takes to fix the first operation when it stops working. If it never takes more than two hours, we want at least two hours' worth of work waiting for station 2. In practice, the buffer would be larger to provide a safety margin.

What happens if station 3 stops working? Should we shut down station 2 to avoid making a pile of inventory in front of station 3? Station 2 makes 12 pieces each hour, while station 3 can make 20 an hour. When station 3 comes back online, it can process the inventory more rapidly than station 2 can replenish it. It will eventually use all of the inventory. Therefore, we should not stop the constraint even if operations downstream stop working.

Thus, operating efficiencies of less than 100 percent are acceptable at any step except the constraint. We never want the constraint to be idle or working at less than 100 percent rate efficiency. (In practice, we may have to accept some downtime for preventive maintenance, and 100 percent OEE may not be achievable.)

Rework and scrap are never desirable, but they are especially bad after the constraint. We previously mentioned opportunity costs or losses of opportunities to make money. Opportunity costs apply only if we can sell everything we can make. (If not, the process has no constraint. The marketplace is the constraint.) Consider the following dollar values in our manufacturing process shown in Figure 2.4.

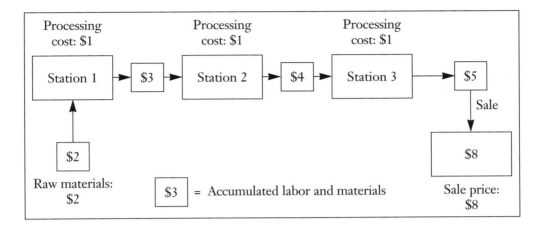

Figure 2.4. Manufacturing costs.

Scrap is usually more costly than rework, because it loses everything that went into the workpiece. For example, a scrap loss after station 1 costs $3 to replace. It costs station 1 $1 to rework a defective item. Figure 2.5 describes a scrap loss after station 1.

The net gain/loss on this event is zero. We spend $3 to make a unit and throw it away because it is scrap. We spend $5 to make another unit, which we sell for $8. $8 in revenue minus $8 in costs is zero. Now, what happens if we rework a unit at station 2? It's easier to understand this by looking at an hour's worth of production. We can normally ship 12 pieces each hour. The profit on each is $3, so an hour's work should earn $36.

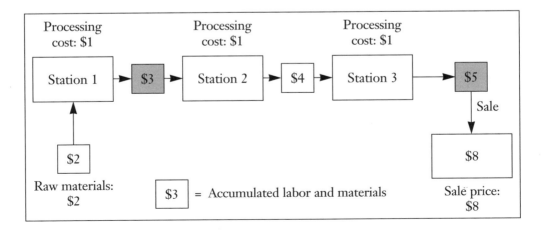

Figure 2.5. Scrap loss after nonconstraint.

The constraint cannot process more than 12 pieces an hour. The rework counts as a piece, since it uses time on the machine. The reworked piece uses two units of capacity, but generates only one sale. The good pieces use the other 10 units of capacity and produce 10 sales. Figure 2.6 shows how rework at the constraint can be worse than scrap before the constraint.

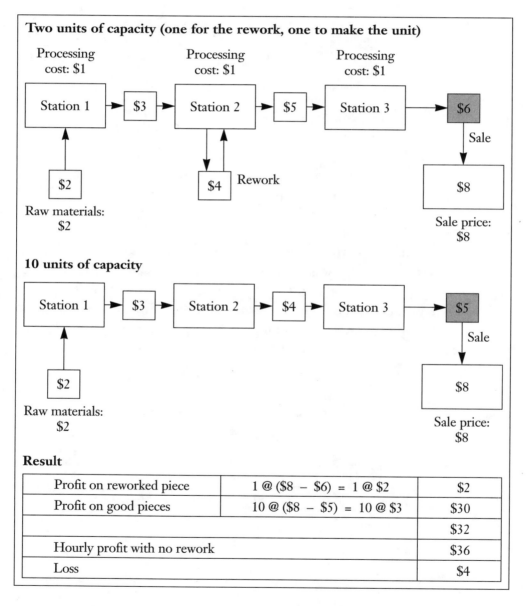

Profit on reworked piece	1 @ ($8 − $6) = 1 @ $2	$2
Profit on good pieces	10 @ ($8 − $5) = 10 @ $3	$30
		$32
Hourly profit with no rework		$36
Loss		$4

Figure 2.6. Rework at constraint.

The $4 loss includes the dollar it cost to rework the part, *plus the loss of the opportunity to earn $3.* The rework tied up a unit of capacity at the constraint. This limited the hour's production to 11 instead of 12. Since the workstation cannot work faster, it can never replace the missing unit. Meanwhile, the cost accounting system reflects only the $1 rework cost. It does not report the opportunity cost of being short a piece.

The scrap loss after station 1 cost only $3, the cost of replacing the unit. Station 1 is not the constraint, and it can replace the missing unit by working faster. This is the key difference. In summary,

1. Operations before the constraint can replace rework or scrap by working faster.

2. Rework or scrap at the constraint cause irrecoverable losses. The constraint cannot replace the missing pieces.

3. Rework after the constraint is recoverable, since the processes can work faster.

4. Scrap after the constraint is irrecoverable, since the constraint cannot replace the losses.

5. The cost accounting system does *not* reflect opportunity costs.

	Rework	Scrap
Before the constraint	Recoverable	Recoverable
At the constraint	Irrecoverable	Irrecoverable
After the constraint	Recoverable	Irrecoverable

These considerations should help set priorities for process improvements. Improvements should usually focus on avoiding scrap or rework at the constraint and avoiding scrap after the constraint.

Check Sheets

Check sheets or tally sheets are simple tools for answering the question, "How often does something happen?" This question could apply to a range of measurements. "How often is the dimension between 3.5 and 3.6 microns?" It could apply to defects or scrap. "How often do contamination defects occur? How often does misalignment (between semiconductor device levels) occur?"

The check sheet is the starting point for a *histogram* or a *Pareto chart*. These are graphical tools for showing how often certain events occur.

Table 2.4 shows rework in a semiconductor plant from October 6, 1990 to October 9, 1990. The manufacturing personnel counted the occurrences of each rework source and recorded it. For readers who do not work in the semiconductor industry, *bridging metal* is a short circuit between the microscopic metal wires on a semiconductor device. *Misalignment* means that successive layers of the device do not line up. It's similar to trying to screw two metal pieces together and finding that the holes don't mate. Semiconductor factories produce the microscopic features with a photolithographic process, which is like photography. Poor development causes poor definition of the microscopic features. Poor contact is another defect, and microscopic contaminants can destroy the product.

This check sheet provides some immediately useful information. First, poor contact accounts for more than half of the reworks. A *Pareto chart* graphically shows that this is the worst problem.

There were five reworks for contamination on October 6. Is this unusually high, when we consider the other days' contamination reworks? An *attribute control chart* tells us if defects, rework, or scrap are unusually high. Attribute data, such as rework, scrap, or defect counts, are whole numbers. Variables or quantitative data, such as measurements, are real numbers. A *multiple attribute control chart* is little different from a check sheet. It tells us if a particular cause is making too many defects or rejections. We will look at Pareto charts and control charts later.

Table 2.4. Check sheet for semiconductor reworks.

Rework cause	October 1990				Total or frequency
	6	7	8	9	
Bridging metal	┼┼┼ I	┼┼┼	┼┼┼ III	┼┼┼ II	26
Misalignment	I	III	III	II	9
Contamination	┼┼┼	I	I	I	8
Poor contact	┼┼┼ ┼┼┼ ┼┼┼	┼┼┼ ┼┼┼	┼┼┼ ┼┼┼ II	┼┼┼ ┼┼┼ ┼┼┼	52
Poor development	II	III	I	I	7
Total reworks	29	22	25	26	102

Histograms

Histograms are bar graphs that show how often something happens. To create a histogram, start by summarizing the totals from the check sheet as follows. Each mark means one occurrence of the rework cause. The length of the line of marks shows the problem's severity graphically. Figure 2.7 shows an example.

Figure 2.8 shows how turning this figure 90° counterclockwise produces a histogram. The horizontal bars become vertical. Each bar's height represents the frequency of the event. A histogram is a graph that shows how often something happens.

Rework cause		Total
Bridging metal	|||||||||||||||||||||||||||||	26
Misalignment	|||||||||	9
Contamination	||||||||	8
Poor contact	|||||||||||||||||||||||||||||||||||||||||||||||||||||	52
Poor development	|||||||	7

Figure 2.7. Check sheet: Problem totals.

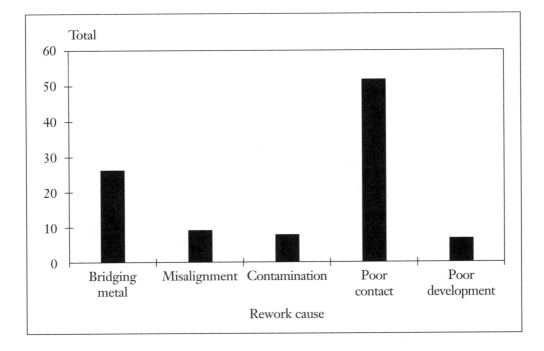

Figure 2.8. Histogram.

Pareto Charts

A Pareto chart is a histogram that displays the events in the order of their frequency. It orders the histogram bars from highest to lowest. Figure 2.9 shows the previous histogram in Pareto chart form. It shows the rework causes in decreasing order of severity.

The Pareto chart is a tool for ranking and sorting alternatives. It helps prioritize manufacturing problems for attention. The Pareto Principle says that most trouble comes from a few causes. The rule of thumb is that 80 percent of the trouble comes from 20 percent of the sources. The 80:20 rule is not exact. In the preceding example, one source (poor contact) caused half the reworks. That is, 20 percent of the sources caused 50 percent of the trouble.

The Pareto Principle supports the "three strikes" laws. These laws seek to identify career criminals who repeatedly commit violent crimes. The Pareto Principle suggests that a few criminals commit most crimes. Focusing prosecution and prison resources on a few offenders should, in principle, prevent most violent crimes.

In manufacturing, defects, rework, and scrap are the "criminals" that vandalize and steal work. Focusing resources and efforts on the few major causes has the biggest payoff.

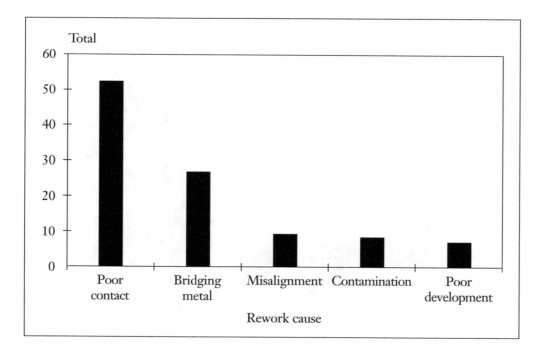

Figure 2.9. Pareto chart.

It is easy to make histograms and Pareto charts with a spreadsheet (such as Corel Quattro Pro, Lotus 1-2-3, or Microsoft Excel). The following data table is from Excel. The DATA menu includes a SORT tool, which sorted the entries in descending order. The next step was to use the Chart Wizard to create a bar chart (Figure 2.9).

Rework cause	Total
Poor contact	52
Bridging metal	26
Misalignment	9
Contamination	8
Poor development	7

Data for a Pareto chart can come from the following sources.

- Use real-time observation.
  - —Record events as they occur. The check sheet or tally sheet is very useful for this.
- Use historical data.
  - —Construct the Pareto chart from historical records.
  - —Be aware, however, that old records have less relevance to the current situation.
- Perform random sampling.
  - —Chapter 3 includes a discussion of random sampling and systematic sampling. The sampling procedure should give each item or workpiece an equal chance of selection.

Histograms—Continued

Histograms can also show the frequency of numerical measurements. These histograms are important in SPC, because they show the measurements' distribution. The following example shows the construction of a data histogram. Doing this by hand is very time-consuming, especially if there are a lot of measurements. You should not have to do this manually. It is, however, very simple for a computer to do this. Microsoft Excel has a histogram tool in its TOOLS—DATA ANALYSIS menu.* Other

*You may need to install this by selecting TOOLS—ADD-INS and checking the analysis pack option.

spreadsheets have similar capabilities. The example's purpose is to show what the computer is doing.

A semiconductor manufacturing process deposits silicon dioxide on wafers. The operation produces the following 50 measurements in angstroms (Å). An angstrom is a very tiny unit of length. It is 1/10 of a nanometer, 1/10000 of a micron, or 1/10,000,000 of a millimeter. The millimeter is the smallest division on a metric ruler. We express atomic sizes in angstroms. The silicon dioxide layer is very thin. Here are the 50 measurements.

1010.0	991.6	998.1	1005.1	983.8
987.4	1006.4	1008.3	996.6	1007.9
1002.3	999.8	1002.3	988.9	1000.9
1008.2	976.0	992.2	992.4	995.0
993.8	989.7	1010.9	996.3	1004.9
1011.4	1003.2	989.5	997.3	1001.3
989.9	1019.2	1006.5	1005.7	1012.9
983.3	1016.8	999.8	988.7	1016.5
990.8	1000.0	994.6	1004.4	1005.5
1011.9	1009.3	990.5	994.3	995.1

To put these data into a histogram, define *cells* for the data. Then use the check sheet technique to count the measurements in each cell. We will sort the data to make them easier to count.

976.0	990.8	996.6	1003.2	1008.3
983.3	991.6	997.3	1004.4	1009.3
983.8	992.2	998.1	1004.9	1010.0
987.4	992.4	999.8	1005.1	1010.9
988.7	993.8	999.8	1005.5	1011.4
988.9	994.3	1000.0	1005.7	1011.9
989.5	994.6	1000.9	1006.4	1012.9
989.7	995.0	1001.3	1006.5	1016.5
989.9	995.1	1002.3	1007.9	1016.8
990.5	996.3	1002.3	1008.2	1019.2

The upper number in the cell is not inclusive, while the lower one is. For example, 1000.0 goes in the cell for 1000 to 1005, not 995 to 1000. (In mathematical terms, the range for cell "995 to 1000" is [995, 1000). This means that the range includes numbers greater than or equal to 995, but less than 1000.) Figure 2.10 shows the data check sheet. Figure 2.11 shows the histogram of the data. Recall that rotating the tally marks 90° counterclock-wise produces a rough histogram. The numbers on the axis are the upper limits of the cells.

The number of cells is a judgment call. Too few or too many cells will yield a histogram that does not show the data distribution well. A good start

(1) Cells	(2) Count	(3) Total
Up to 980	I	1
980 to 985	II	2
985 to 990	IIIIII	6
990 to 995	IIIIIIII	8
995 to 1000	IIIIIIII	8
1000 to 1005	IIIIIIII	8
1005 to 1010	IIIIIIIII	9
1010 to 1015	IIIII	5
1015 or more	III	3

Figure 2.10. Data check sheet.

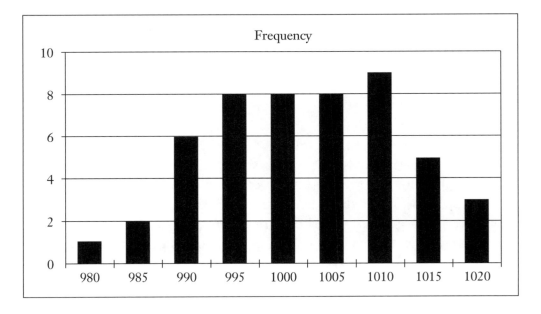

Figure 2.11. Data histogram.

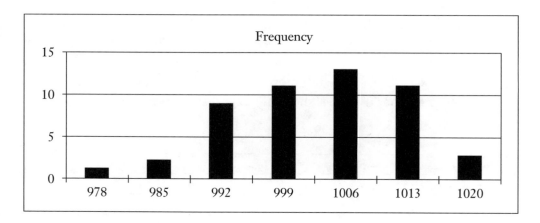

Figure 2.12. Data histogram after cell change.

is to use the square root of the data count. Here, the square root of 50 is slightly more than seven. Since it is easy to count the data if the cell width is five, Figure 2.11 has nine cells. Figure 2.12 shows a histogram of the same data, using seven cells and cell widths of seven.

Why would we want to display data in a histogram? The SPC section will show that data from most manufacturing processes follows a normal distribution or bell curve. The histogram's shape should be similar to a bell, and the histogram in Figure 2.12 is roughly bell-shaped. The SPC chapter shows histograms that use more measurements. The bell shape is much clearer in those.

Process Flowcharts

A flow diagram shows the steps in a manufacturing process. It is useful for understanding the process and for identifying critical operations. Hradesky (1988, 27–29) recommends the symbols shown in Table 2.5 for process flowcharting. The choice of symbols, however, is not critical.

A refinement of the chart can show where materials enter the process. Raw materials or subassemblies can affect product quality or productivity. Another refinement (Hradesky 1988, 173–175) shows an operation block for each workstation that can perform the operation. The cause-and-effect (fishbone) diagram is a useful tool for problem solving. Two standard branches for this diagram are machines (equipment) and materials. Showing the workstations and material flows helps the manufacturing team identify potential problems with machines and materials.

Table 2.5. Flowchart symbols.

Operation (production or activity	◯	Flow of work	⟶
100% inspection, measurement, or testing	▢	In-process inspection or measurement, with a control chart (operation #1)	⊡〰️ ①⟶②
Inspection, measurement, or testing with a control chart	〰️	Operation followed by 100% inspection or measurement	◯⟶▢
Storage	▽	Operation followed by 100% inspection or measurement, with a control chart	⊡〰️ ◯⟶▢

Consider the following assembly process in a semiconductor factory. A die, chip, or pellet is the actual transistor. Here is how it gets into the familiar three-legged transistor package. A die attachment machine solders the die to a stem. A wire bonder solders two wires to connection points at the top of the die. Another machine places a protective cap over the assembly. Finally, the package undergoes an electrical test (Figure 2.13).

Figure 2.14 shows a flowchart for this process. There are two die bonding workstations, two wire bonding stations, and one capping station.

Now suppose we get stoppages in the die bonding operation. Figure 2.13 shows only where the die bonding operation fits into the manufacturing process. Figure 2.14 shows that there are two die bonders, and that they use die, stems, and solder. An obvious question is whether both bonders are experiencing stoppages. If the answer is no, one workstation is at fault and needs repair. If both bonders have stoppages, the materials are probably at fault. It is unlikely that both bonders would have the same mechanical malfunction simultaneously. Suppose that both bonders are having the problem, and that the company recently started buying stems from a new vendor. What is the most likely problem? (The stems must meet a size specification so the bonder can handle them properly. If they are not the right size, they will jam the machinery.)

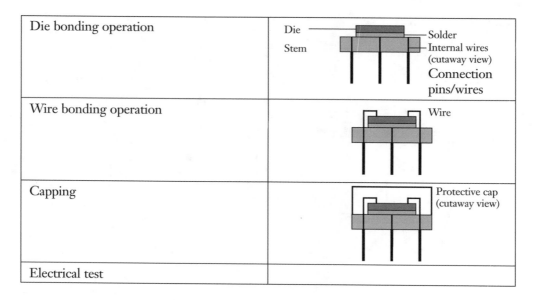

Figure 2.13. Semiconductor package assembly.

Critical Processes

According to Frederick II, King of Prussia, "One who tries to protect every-thing ends up protecting nothing." His comment on military operations also applies to manufacturing operations. It is a common mistake to disperse the manufacturing team's attention among every operation. Some companies see SPC as a magic Japanese success secret. (SPC was actually an American invention, which the Japanese adopted after the Second World War.) By applying SPC to everything, companies think they will automatically improve quality and productivity. No one, however, can watch all the charts. The charts cover the walls and look impressive, but serve no other purpose. Hradesky (1988) calls these charts "wallpaper."

It is far more effective to control the vital operations that affect productiv-ity and quality. To control a process effectively, the manufacturing team must identify the critical operations. Harris Semiconductor defines these operations as follows: A *critical process* or *critical node* is an operation that can significantly affect yield, productivity, reliability, or performance. A process is also critical if it affects a customer-defined specification or special characteristic.

This means that, if the critical node does not perform properly, one or more of the following may happen.

1. There will be more rework or scrap.

2. Productivity will be lower. The manufacturing team will get fewer pieces for the same effort.

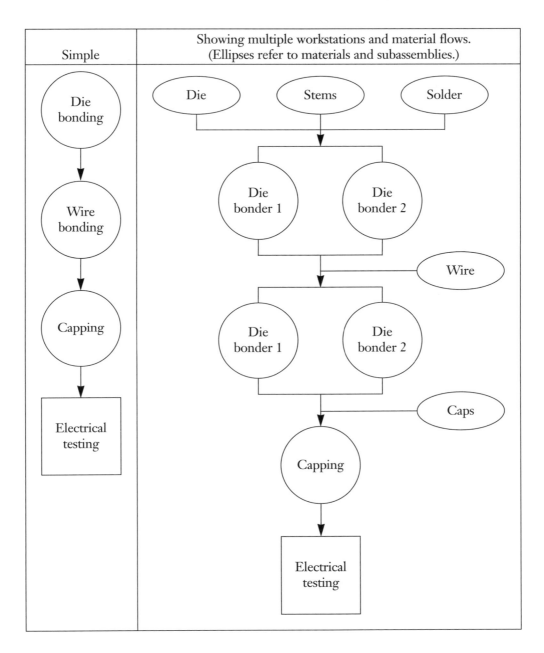

Figure 2.14. Flowcharts for semiconductor package assembly.

3. Defects in the products will make them less reliable. Although they pass final inspection, they are more likely to fail in service.

4. Defects in the products will lower their quality (performance). They will not meet the customer's needs as well. Their fitness for use will be lower.

5. The products will not meet the customer's specifications.

Cost of Quality

Cost of quality (COQ) analysis quantifies the cost of poor quality. It defines quality problems, or opportunities for improvement, in the language of money. This is the language of upper management, and it is useful for gaining management support for quality improvement projects. It is also useful for quantifying these projects' benefits.

COQ analysis fits with the process flowchart. Each activity in the flowchart adds value to the product, affects its quality, or both. If we know the cost of each activity, we can tabulate it. Table 2.6 shows the costs of quality. See chapter 4 of Juran and Gryna (1988) or chapter 7 of Feigenbaum (1991) for more details. In the flowsheet for semiconductor assembly, die bonding, wire bonding, and capping are required operations. The electrical test is an appraisal.

The COQ is A + P + F. This is the cost of poor quality or the cost of avoiding poor quality. That is, we must spend money on prevention and appraisal to avoid making or selling bad products. The *cost of quality percentage* (COQ%) is $\frac{A + P + F}{R + A + P + F} \times 100\%$. This is the ratio of the quality costs to the total manufacturing costs.*

The following list shows the costs of (poor) quality in increasing order of usual desirability. Prevention is usually less expensive than appraisal and internal failure, which are better than external failure.

1. Get the disease, don't treat it, and suffer from it. Thus, External failure = Angry customers

2. Diagnose the disease and treat it. Thus, Appraisal = Diagnosis and Internal failure (rework, scrap) = Treatment

3. Prevent the disease. Thus, Prevention = Vaccination

---

*The philosophy is that required activities add value for the customer, while prevention, appraisal, and failure activities do not. External failures actually create negative value for the customer. After the product leaves the factory, we should examine intermediaries like retailers and dealers. Do they add any value for the customer?

Table 2.6. Costs of quality.

Category	Symbol	Description
Required	R	Required operations add value to the product.
Appraisal	A	Appraisals are inspections and tests whose purpose is to assure outgoing quality. Appraisal activities *detect* nonconforming items and cull them for rework or scrap. They keep them from reaching the customer, but do not prevent them. Appraisals include • Acceptance sampling • Incoming inspection • Final inspection or test
Prevention	P	Preventive activities *prevent* rework, scrap, and other failures. Activities include • Process control, including SPC • DFM or designing quality into the product; DFM means designing for ease of manufacture • Preventive maintenance or TPM • Most ISO 9000 activities
Failure	F	Internal failure costs include the costs of • Rework • Scrap • Rectification or detailing of lots that fail an acceptance sampling test • Reinspection or retest Remember that accounting costs may not reflect the full cost of rework and scrap. They may cause opportunity costs or losses of opportunities to make money.
		External failure is usually the worst event that can happen. It means the product fails after the customer has received it. Its costs include • Warranty costs • Returns • Customer dissatisfaction, which is an intangible cost

Cause-and-Effect (Fishbone) Diagrams

A *cause-and-effect diagram* helps a manufacturing team identify potential problem sources. It is a systematic tool for helping a group organize its thoughts.

The die bonder example provided a verbal or narrative discussion of the possible reasons for stoppages. A cause-and-effect diagram systematically organizes these causes. Another name for this diagram is the *fishbone diagram*, since it looks like a fish skeleton. Some people call it an *Ishikawa diagram*, after its inventor.

We will construct a cause-and-effect diagram for the die bonder problem. The problem cause goes in the "fish's head." The diagram's major branches, or the fish's major bones, are problem categories. They prompt the team to ask questions like, "How could the working environment cause this problem?" "How could the materials cause stoppages?" Contino (1987) uses the categories in Table 2.7.

Table 2.7. Major problem categories, cause-and-effect diagram.

Methods	The procedure or operating instructions • These must be clear and easy to understand
Measurements	"If you can't measure it, you can't control it." • Gage accuracy or calibration • Gage precision, or reproducibility and repeatability (R&R)
Manpower	Personnel* or human factors include training and motivation • Motivation problems could include performance measurements that drive undesirable behavior.
Materials	Raw materials and subassemblies

Machines	Equipment and tools
	    There is sometimes a question about what is a material and what is a machine. A material becomes part of the product or is a consumable item like a chemical. Materials appear on the bill of materials (BOM). The cost accounting system treats them as expenses.
	    Machines are durable items, and the cost accounting system often depreciates them. It may, however, treat inexpensive tools as expenses.

Environment	• Temperature
	• Humidity
	In microelectronics manufacturing, add
	    • Noise and vibration
	    • Particulate contamination (this is why we have cleanrooms)
	    • ESD
	Food processing, pharmaceutical, and medical activities must consider
	    • Bacterial contamination

Personnel or *people* would better describe the modern workforce. *Manpower*, however, helps users remember the problem classes as "5 Ms and an E." (Some people refer to the environment as the *medium*, which provides "six Ms.")

The process flowchart showed three materials or subassemblies that enter the die bonder. These are the die, electronic chip, or pellet; the stem; and the solder. The die and the solder do not enter the part of the machine where the jam occurs. The stem, however, does. The presence of the stem in the diagram prompts the question, "How could the stem jam the machine?" Perhaps the stem is the wrong size. Meanwhile, what could go wrong with the machine itself? Does it need PM that it is not receiving? Could clutter in the work area provide foreign objects that can jam the machine? Could temperature variations make parts or machinery expand or contract, thus causing stoppages? Temperature and humidity are standard considerations in many manufacturing operations. Are operators making mistakes because of poor training? Are they making mistakes because the operating instructions are unclear? Are they working with outdated instructions or specifications? Figure 2.15 shows the cause-and-effect diagram.

The diagram's branching feature prompts the group to look for underlying causes. For example, the entry "stems" prompts the team to think about why the stems could jam the machine. Could they be the wrong size, or did something bend them out of shape? Either of these problems could make them jam the bonder. These two possibilities should make the team ask further questions. What could damage the stems and bend them out of shape? Do people or machines handle them? Can damage occur during transportation? Why would they be the wrong size? Is the supplier meeting

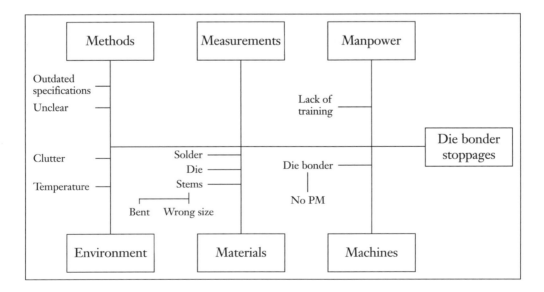

Figure 2.15. Cause-and-effect diagram, die bonder stoppages.

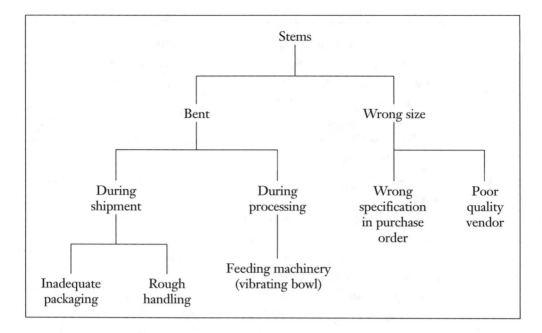

Figure 2.16. Branching on cause-and-effect diagram.

specifications? Figure 2.16 shows further branching of the "stem" item. (The vibrating bowl aligns the stems for the die bonder's conveyor mechanism.)

Readers should be able to name a couple of activities that address the entries "clutter" and "no PM" on Figure 2.15. We will also look at a program that deals with "outdated specifications."

Problem	Addressed by
Clutter	5S-CANDO
No PM	TPM
Outdated specifications	ISO 9000

Case Study: Wafer Breakage

Semiconductor manufacturing involves coating silicon wafers with photosensitive films. These films, or photoresists, allow the creation of microscopic patterns on the wafers. The silicon wafers are brittle, and they require careful handling. A manufacturing team was experiencing wafer breakage during this process.

The team called a meeting and set up wafer checkpoints at five process steps. The team used a check sheet to collect data for one month. Here are the results. First, Figure 2.17 shows a Pareto chart of the breakage causes.

Defect	Abbreviation	Count
Wafers broken at spin coating step 1	Broken, coater 1	29
Wafers broken at spin coating step 2	Broken, coater 2	4
Operator error	Operator	5
Wafers broken during deposition	Broken, dep	6
Miscellaneous	Misc	1

Coating operation #1 is clearly the source of most of the breakage. The team called a meeting and developed a cause-and-effect diagram. Here is some information about the spin coating operation. The operator had to pick up the wafer with metal tweezers. (Touching a silicon wafer with one's hands, even with gloves, causes unacceptable contamination.) Squeezing the wafer too hard can break it. Squeezing it too loosely can allow it to fall out of the tweezers. In the 1960s, wafers were 1″ (25.4 mm) in diameter and very thin.

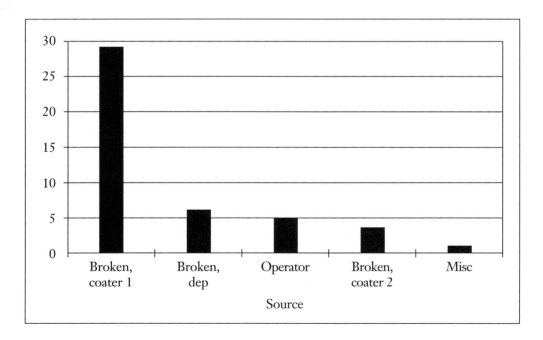

Figure 2.17. Pareto chart of wafer breakage sources.

As manufacturing technology evolved, wafers became larger, thicker, and heavier. Manufacturers routinely use 5″ and 6″ (125 mm, 150 mm) wafers. Harris Semiconductor's Mountaintop plant is adding equipment that will handle 8″ (200 mm) wafers. These wafers are very heavy and very expensive. Think about how handling considerations might have changed over the years.

The operator places the wafer on a vacuum chuck. After placement of liquid photoresist on the wafer, the chuck spins the wafer at 5000 revolutions per minute (rpm). The centrifugal force causes the solution to spread across the wafer and form a uniform film. Centrifugal force is proportional to the square of the spin speed.

Next, the wafer went on a 110°C (230°F) hot plate. This step evaporates the solvent from the photoresist to leave a film. The team suspected that thermal shock could break the wafer.

The team constructed a cause-and-effect diagram, which appears in Figure 2.18. (Measurements were not applicable. Removing this item provides more room for methods.)

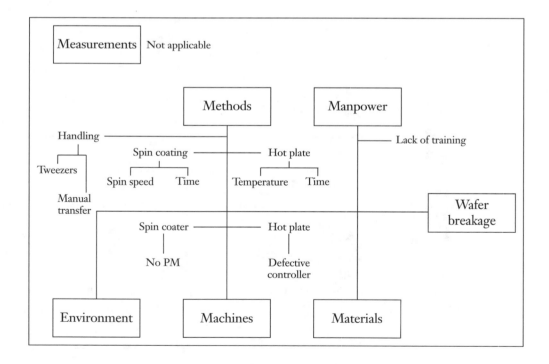

Figure 2.18. Cause-and-effect diagram, wafer breakage.

The spin speed and time appear under spin coating for methods because they are in the process recipe. "No PM" appears under the spin coater for machines. Lack of preventive maintenance could cause the spin coater to malfunction. The wrong spin speed or time would be a method problem. If so, the machine does what we want, but what we want is bad for the product. Similarly, temperature and time are methods for the hot plate bake. The hot plate does what we tell it to do. If the hot plate's controller is defective, the hot plate doesn't do what we want, and this is a machine problem.

The team's experience with the process, and further investigation, allowed the team to rule out some of the items. Table 2.8 summarizes the team's decisions.

Table 2.8. Manufacturing team's actions.

Action	Reason
Purchase a wafer transfer machine to replace manual handling with tweezers.	When wafers were smaller, a very light grip was adequate to hold them. Tweezers were acceptable for small, light wafers. The team questioned whether tweezers are appropriate for handling large wafers. The user must apply more pressure to hold them. Squeezing too hard will break the wafer. Squeezing too lightly will let the wafer slip out of the tweezer, fall on the floor, and break.
Replace tweezers with vacuum wands in other operations.	Same as above. A vacuum wand is a flat tool that holds the back of the wafer by suction. A large vacuum wand will hold a large wafer.
Ask engineering to design an experiment on coater spin speed and hot plate temperature.	As the wafer diameter increases, centrifugal force also increases. (See Figure 2.19.) The g-force on the outer rim of a 4″ wafer is twice what it is on a 2″ wafer at the same spin speed. Could this be sufficient to break the wafer? Thermal shock from putting a cold wafer on a hot plate could fracture the wafer. Results 　• Change the spin speed from 5000 to 2500 rpm, and raise the spin time to yield the same coating thickness. Spin time will be about four times as long at half the spin speed. 　• Lower the temperature to 80°C.

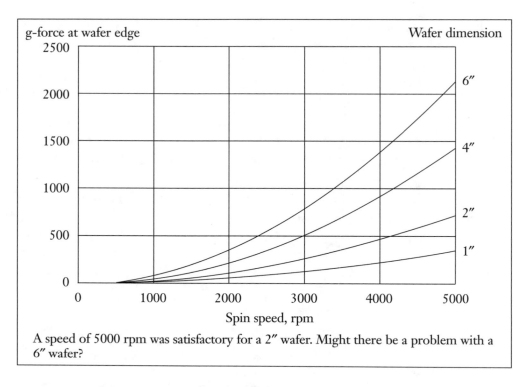

Figure 2.19. Centrifugal force at wafer edge during spin coating.

After the changes went into effect, the team collected another month's worth of data. Here are the results; Figure 2.20 shows a Pareto chart.

Defect	Abbreviation	Count
Wafers broken during deposition	Broken, dep	3
Wafers broken at spin coater 2	Broken, coater 2	2
Wafers broken at spin coater 1	Broken, coater 1	0
Operator error	Operator	0
Miscellaneous	Misc	0

The team was very successful in reducing the breakage. Why is "operator error" now zero? When the operator broke a wafer with the tweezers, or dropped it, people called it an operator error, but it really wasn't. The problem's root cause was inappropriate tooling for the job. Once the operators had the right tools, the problem went away.

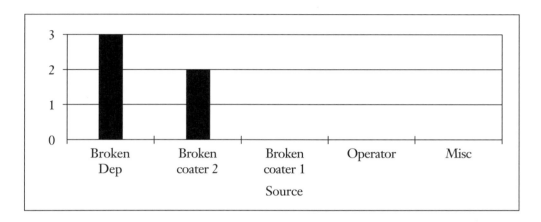

Figure 2.20. Pareto chart of wafer breakage sources after changes.

Team-Oriented Problem Solving: Eight Disciplines

Ford Motor Company's TOPS-8D is a systematic approach for solving problems. It contains all the elements of the plan-do-check-act (PDCA) cycle. It is particularly suitable for self-directed work teams.

Juran and Gryna (1988, 10.25–10.26) describe the PDCA cycle for improving a process. Here is their PDCA cycle.

1. Plan

   a. Assemble a project team.

      1. The team may already exist as a manufacturing team or self-directed work team.

      2. Project teams should be cross-functional. Ask appropriate people outside the department to participate. Include the people who have expertise or skills that can help solve the problem. These may be

         a. Process engineers

         b. Product engineers

         c. Quality engineers

         d. Purchasing or marketing personnel

         e. Others

   b. Identify the problem or improvement opportunity.

      1. How does the process' actual performance differ from its desired performance?

c. Plan an experiment to reduce or remove the gap between the desired and actual performance.

2. Do
   a. Perform the experiment.
   b. Collect data.

3. Check
   a. Assess the experiment's results. How did the change affect the process?

4. Act
   a. If the change did not produce desirable results, return to step 1.
   b. If the change improved the process, make it permanent.
      1. Hold the gains by going to a control cycle (Juran and Gryna 1988, 10.26).
      2. SPC is a feedback control cycle.
      3. Nonstatistical controls may, however, be effective. It depends on the process.

Here are the key instructions for Ford Motor Company's TOPS-8D method.

1. Use a team approach.
   a. The team should have a champion.
      1. This person should have ownership of the problem. He or she should have the resources and authority to implement the team's decisions.
      2. The champion is not necessarily the team leader.
   b. Other roles include a leader, recorder, and facilitator.

2. Develop a working definition for the problem.
   a. Compare what should have happened to what actually happened. As in PDCA, how does the process' actual performance differ from its desired performance?
      1. Be sure to clearly identify and define the problem. Juran and Gryna (1988, 22.35) say that problem statements are often ambiguous or imprecise. This makes it hard to solve the problem or improve the process. Make sure that everyone is trying to fix the same problem!

2. Ask "what, where, when, and how big." What is the problem? Where is it seen or detected? When (in time or in the process flow) is it seen? How big is it, or how many pieces does it affect? Quantify the problem's severity.

3. An "is/is not" description is helpful. Knowing what the problem isn't, as well as what it is, helps the team locate its source. Combine the "is/is not" question with "what, where, when, and how big."

b. Where does detection or identification of the defect happen?

1. Is it detectable at its source?

2. Do we have to wait for an inspection, measurement, or further processing before we can detect it?

3. The process flow diagram is useful here.

c. What is the defect's source?

1. Which process step generates it?

2. Is the defect traceable to a particular workstation?

3. Contain the problem. This means stopping it from causing damage while the team is trying to fix it. (This assumes that a previously acceptable process has started producing unacceptable results. A sudden increase in rejects is an example.)

a. Containment is similar to quarantine. Quarantine does not cure a disease, but it keeps it from spreading.

b. Find and segregate nonconforming products to prevent their shipment to customers.

c. Shut down manufacturing equipment that is making bad product.

d. Make sure the containment action is effective.

4. Identify the problem's root cause.

a. This is like diagnosing the disease.

b. Tools for doing this include the check sheet, Pareto chart, and cause-and-effect diagram.

c. Design of experiments (DOE or DOX) is a quantitative tool for investigating a problem or trying an improvement. It requires consultation with an industrial statistician or an engineer who knows statistics.

1. A statistically designed experiment can provide a confidence level that a change improved the process.

5. Select a permanent correction for the root cause, and make sure that it works. This is like curing a disease.

6. Carry out the permanent correction. Monitor the process to make sure it is effective. This is like watching a patient to make sure the cure worked.

7. Prevent the problem from coming back. This means holding the gains (step 4 of PDCA).

8. Recognize the team's accomplishments. Each company has its own procedures for doing this.

"Is/Is Not" Analysis

Here is an example of how defining what a problem isn't, as well as what it is, can help locate its source. Refer again to the die bonder stoppage problem and the process flow diagram (Figure 2.14). Table 2.9 shows an "is/is not" description of the problem.

The entries in the "where" row help us rule out a die bonder problem as a root cause. (It's unlikely that both stations would have the same problem at the same time.) The problem does not affect the wire bonder or the capper, so we can rule out these steps. Similarly, the caps, wire, and solder are not possible root causes. This analysis points to either the die or the stem. Since the stoppage does not involve the die handling mechanism, it's probably the stem.

Now suppose the "is/is not" description looks like Table 2.10. What is the likely root cause?

The "is/is not" table provides a convenient, qualitative way to test theories about the root cause. For example, the theories might be entries on a cause-and-effect diagram. Each possible root cause must be consistent with most (preferably all) entries in the "is/is not" table. Cross-tabulate the theories

Table 2.9. "Is/is not" description of die bonder stoppage.

	Is	Is not
What	Die bonder, stoppage	Wire bonder or capper stoppage A soldering problem
Where	**Both wire bonding stations** Stem handling mechanism	**Only one station** Die handling mechanism
When	Intermittent	Consistent
How big		

against the is/is not entries as follows. Enter the following symbols in the table, as shown in Table 2.11. Use the information from Table 2.9.

+   Explains both the "is" and "is not"
–   Is inconsistent with the "is" or the "is not" description
?   Need more information

"Problem with the stems" is the only root cause in the table that has no – symbols. It is the only explanation that is consistent with all the information. This does not prove that stems are the problem source, but it tells us where to look.

Table 2.10. Alternate "is/is not" description of die bonder stoppage.

	Is	Is not
What	Die bonder, stoppage	Wire bonder or capper stoppage A soldering problem
Where	**Only one station** Stem handling mechanism	**Both wire bonding stations** Die handling mechanism
When	Intermittent	Consistent
How big		

Table 2.11. Testing possible root causes of die bonder stoppages.

	Is	Is not	Possible root cause		
			Problem with the die bonder	Problem with the stems	Problem with the die
What	Die bonder, stoppage	Wire bonder or capper stoppage	+	+	+
Where	Both wire bonding stations	Only one station	–	+	+
	Stem handling mechanism	Die handling mechanism	+	+	–
When	Intermittent	Consistent	+	+	+
How big					

ISO 9000

Murphy's Law says that anything that can go wrong, will. ISO 9000 is a system for making sure that everything will go right.

ISO 9000 is a set of international standards for quality management systems. It applies to management and process controls for assuring quality. Companies that meet these standards can receive ISO 9000 registration from accredited registrars. Many industrial customers are starting to require their suppliers to achieve ISO 9000 certification. The standards include ISO 9002, which is for production, installation, and servicing. ISO 9001 includes all the elements of ISO 9002 and adds standards for product design.

The automotive QS-9000 quality standards are extensions of the ISO 9000 standards. An organization that qualifies for QS-9000 certification also meets ISO 9000 standards. Meeting ISO 9000 standards, however, is not sufficient for QS-9000 certification.

How does ISO 9000 help us do our jobs? How does it help us deliver quality products and services consistently? ISO 9000 systematically guides an assessment of our quality management system and process controls. It helps us look at how we make the product or deliver the service. It makes us ask, "What can go wrong, and how can we change the system to prevent it from doing so?"

Some companies want to "get the ISO 9000 certificate so we can do business in Europe." They view the certification process as a costly annoyance. This is an error. The systematic assessment of the quality management system can improve productivity and quality and help the company make money (Scotto 1996). This section will not discuss the details of ISO 9000, but will give examples of its applications.

Documentation Control

Primitive tribes transmitted their lore orally, since no one could read or write. This oral tradition may have been the origin of poetry. The rhyming verses helped the storyteller or speaker remember what to say. This is not, however, how we want to run a manufacturing process. There must be written instructions and specifications to make sure that everyone does the job the same way.

Document changes are a common trouble source. Suppose that an operating instruction says to set a furnace temperature to 350°C. A change in materials requires operation at 370°C, so the process engineer revises the operating instruction. What happens if he or she delivers the revision to the manufacturing area, but an old copy remains there? Someone may pick up the old instruction and set the oven to the wrong temperature.

ANSI/ISO/ASQC Q9002-1994's section 4.5 "Document and Data Control" requires procedures to prevent this from happening. It specifically calls for controls to "preclude the use of invalid and/or obsolete documents" (ANSI/ISO/ASQC 1994). Thus,

- There must be procedures for removing obsolete documents from manufacturing areas.
    - —A specific person may be responsible for replacing obsolete documents with new ones.
    - —Electronic databases can assure that only the current instructions are available. There must, however, be a computer monitor near each workstation.
- Everyone who performs a job must know when the instructions change.
    - —Harris Semiconductor uses a revision sign-off log. After reading the new instruction, each operator signs the revision log. Other companies may have different procedures.

Product Identification and Traceability

Chapter 3 will discuss why archery was inherently superior to gunnery until the nineteenth century. Armies of the fifteenth and sixteenth centuries did not adopt guns because they were superior to bows, but because guns were easy to use. Bows were better weapons, but it took years of training and practice to make a good archer. The nineteenth century saw the invention of breechloading rifles, which are easy to use and are more effective than bows.

One of the bow's advantages was traceability. The archer could watch his arrows in flight and change his aim accordingly. If something was wrong with his bow, he would know it immediately. The archer's product—his shot—was traceable. He knew which arrows were coming from his bow, and which were coming from his neighbors'. The firearms of that era, like matchlock muskets, offered no such advantage. The bullet was invisible in flight and provided no feedback to the gunner.

If our process is making defective product, we need to know from where it is coming. Do all the bad parts come from one workstation or one material lot? Traceability is also a requirement for effective SPC. For example, each workstation should have its own control chart. To put a measurement on the right chart, we must know which workstation produced it. (This assumes that measurement or inspection happens after the part leaves the workstation, and not at the workstation.)

Process Control

Process control includes the following:

- Control of the working environment
  - —Temperature: Temperature affects spin coating processes (semiconductor industry), painting, and other coating operations
  - —Humidity
  - —Particulate contamination (semiconductor industry)
  - —Noise and vibration
  - —Electrostatic discharge (semiconductor industry)
  - —Bacterial contamination (food processing, pharmaceutical industries)
- Control of the manufacturing process
  - —SPC
  - —Automatic feedback process control
- Preventive maintenance
  - —TPM

Inspection, Testing, Calibration, and Control

Sections 4.10 and 4.11 of ANSI/ISO/ASQC Q9002-1994 govern inspection, testing, and control of inspection and testing equipment. Section 4.12 covers the product's inspection and test status. These sections' principal goal is to make sure that

- Products that require inspection or testing actually receive them. This applies to incoming materials and outgoing product. Lot travelers, lot tickets, or routings show the inspection or test status.

- Gages (measuring equipment) receive the necessary calibration. This means comparing the gage's measurement to a standard. If the measurement does not match the standard, we must adjust the gage until it does. *Reconditioning* is another word for calibration.

Suppose that we discover that a gage is out of calibration on April 1. It was in calibration on March 1. There is a quality exposure on every part that went through that gage between March 1 and April 1. (We don't know when the gage went out of calibration.)

Each gage should have a sticker that shows the date of its last calibration and the due date for the next one. Gages that do not require calibration should have stickers saying, "No calibration required." This shows that the manufacturing team is aware of the gage and has judged that it does not need calibration. This consideration applies to gages that measure product or whose performance affects product quality. Room clocks, for example, do not require stickers.

Uncalibrated gages should not be present in the manufacturing area if someone could use them to measure product. It does not matter that no one plans to use these gages on product.*

Calibration is similar to what police and detectives call *chain of custody*. Suppose a gage is in calibration on March 1, but is out of calibration on April 1. Everything that went through it between those dates is suspect. We know that it went out of calibration somewhere between those dates, but not exactly when.

Suppose that a vehicle may contain crime evidence. The police have it in custody, but someone manages to break into it. This compromises the reliability of any evidence the police may collect from it afterward. Similarly, we cannot rely on the quality of product that went through a gage whose calibration is questionable. A lawyer does not want to give questionable evidence to a jury, and we don't want to give questionable product to a customer.

Control of Nonconforming Product

In one of C.S. Forester's *Horatio Hornblower* stories, an English warship receives barrels of spoiled meat. Hornblower decides to exchange the barrels for good ones, but he suspects that the supplier may sell the bad ones to another ship. If the other ship's captain is in a hurry to sail, he won't find out until it is too late. Hornblower therefore orders two midshipmen to write "CONDEMNED" on each barrel with a hot iron. This is an example of *segregation of nonconforming product*.

Today, few people want to use or sell nonconforming material. (The unscrupulous ones who do usually don't stay in business very long.) There is, however, a chance of doing it accidentally. Suppose someone in our incoming inspection department discovers a problem with a batch of incoming material. He or she puts it aside instead of releasing it to the factory, but doesn't mark it. Someone from the factory needs this material and sees the package in the inspection room. He or she may pick it up and take it into the factory. In the production area, someone may take rejects out of a box and put them aside. If they are not marked as rejects, someone else may pick them up and use them.

---

*This concept goes back to the Talmud. "A person is forbidden to keep in his house a measure smaller or larger [than the nominal capacity] even if [it is used as] a urine tub" (Baba Bathra, quoted in Juran 1995, p. 46). Although the measure, or gage, is not intended for trade, someone may use it for that purpose by mistake.

The Talmud also calls for scheduled calibration. "A shopkeeper must clean his measures twice a week, wipe his weights once a week and cleanse the scales after every weighing" (Baba Bathra, V:88a, quoted in Juran 1995, p. 46).

Section 4.13 of the ISO 9002 standard requires suppliers to prevent accidental use of nonconforming materials. Methods for doing this include

- Rework
- Acceptance by the customer, perhaps at a lower price (such as clothing manufacturers selling nonconforming garments as seconds)
- Downgrading
- Scrap

The manufacturer must identify and segregate nonconforming items to prevent their accidental use. A bright orange REJECTED sticker can identify nonconforming materials or parts. A reject rack, or lockable reject cage, can hold nonconforming pieces.

Here is an example of how an electrical tester prevents bad pieces from mixing with good ones. The tester sorts the parts into bins according to the parts' electrical characteristics. Rejects go into the reject bin (Figure 2.21).

- The arm's idle position is over the reject bin. If the arm stops working, it will not put everything into a bin for shippable product.
- The tester counts the nonconforming pieces. This count must match the quantity in the reject bin. If it doesn't, we must suspect that bad ones went into the wrong bins.

Storage, Packaging, Handling, and Delivery

Readers of Tom Clancy's *Red Storm Rising* and *Debt of Honor* will be familiar with this issue. In *Red Storm Rising*, the Soviets transport antiaircraft missiles by ship to Iceland. Seawater (or salt spray) damages some missiles. While the

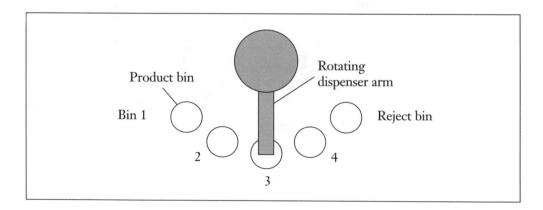

Figure 2.21. Sorting of parts by electrical tester.

Russians designed their naval missiles to resist saltwater, they didn't do this for their army missiles. In *Debt of Honor*, seawater damages some automobile gasoline tanks from Japan. Section 4.15 of the ISO 9002 standard would have been helpful in these situations.

Is the product perishable? Does it have a shelf life? Is it vulnerable to damage by heat, shock, or electrostatic discharge? Here are some practices for addressing these issues.

- Semiconductor and microelectronic products require special packaging to protect them from electrostatic discharge.

- The factory must use materials before their expiration dates.

  —Foods in supermarkets often have "sell by" dates.

  —Recall the option of selling nonconforming pieces at a lower price. Supermarkets often drop the price of fish on the expiration date.

  —Nonprescription drugs also have expiration dates.

- Some chemicals require refrigeration or must not exceed certain temperatures.

  —There are temperature-sensitive stickers for application to packages. The sticker undergoes a permanent color change if the temperature exceeds the specification.

  —Does the supermarket's refrigerator shelf keep the items in front as cold as the items in back? One of the authors usually selects items from behind the front row.

- Is the product sensitive to shock? Will violating the "this side up" instruction damage it?

  —There are tipping indicators that show whether a package has been mishandled.

Other Provisions

Other sections of the ISO 9002 standard include the following:

1. Management responsibility (Section 4.1). Management is responsible for developing and implementing a quality policy.

2. Quality system (Section 4.2). The factory's quality manual provides the basic requirements for the quality system. All work instructions and specifications must support, or conform to, the quality manual.

3. Review of contracts, such as with suppliers or subcontractors (Section 4.3).

4. Control of product design (ANSI/ISO/ASQC Q9001-1994, Section 4.4). This is part of ISO 9001, but not ISO 9002. Manufacturers that do not design products don't have to worry about this section.

5. Assurance that purchases from suppliers meet specifications (Section 4.6).

6. Control of products supplied by the customer (Section 4.7).

7. Control and retention of quality records (Section 4.16).

8. Internal quality audits (Section 4.17). The organization audits its own activities to make sure they follow ISO 9000 requirements.

9. Training (Section 4.18). The organization makes sure that people have the right training for their jobs. Some occupations, like driving a truck or forklift, or doing electrical work, require licenses.

10. Servicing (4.19).

11. Statistical techniques (4.20).

Major Problem Areas

The Pareto Principle applies to ISO 9000. A few program elements are the source of most of the problems. Robert M. Bakker of Entala, Inc. (Grand Rapids, Michigan) said that many companies that are seeking QS-9000 registration run into trouble with the following elements ("Why companies fail audits" 1996).

- Documentation control (element 4.5): "little yellow sticky notes with work instructions stuck to documents or machinery won't cut it."

- Inspection and testing (element 4.10).

- Control of inspection, measuring, and test equipment (element 4.11): "there's often at least one gage with no record of calibration."

Problems

Problem 1. The following defect counts occurred at a pre-probe silicon wafer inspection. Make a Pareto chart of the data.

Defect	Abbreviation	Count
Blisters	BLIS	4
Discoloration	DISC	52
Oxide holes	OX	14
Damage	DAM	163

Problem 2. The following defect counts occurred at a pre-probe silicon wafer inspection. Make a Pareto chart of the data.

Defect	Abbreviation	Count
Blisters	BLIS	31
Discoloration	DISC	2
Oxide holes	OX	20
Damage	DAM	5
Miscellaneous	MISC	2

Problem 3. There is a problem with scratches on silicon wafers. The people investigating the problem counted scratches at the workstations in the manufacturing process. Make a Pareto chart of the data. Where should the manufacturing team focus its attention?

Operation	Abbreviation	Scratches
Aluminum etching and stripping	Etch	94
Silicon nitride deposition	NitD	22
Silicon nitride etching	NitE	97
Back grinding	Grind	2
Ion implantation	Imp	11
Backside metal deposition	BMet	23
Wafer firing	Fire	0
Wafer probing (electrical test)	Prob	34

Problem 4. A factory has three shifts that work 24 hours a day. Compute the OEE for the following machine. The machine is not at a constraint operation. Are the partial loads or idle time problems? What about the rework?

  a. The machine is available 24 hours a day.

  b. The machine operates, on average, 18 hours a day.

  c. The machine can handle 50 pieces. The usual load is 40 pieces.

  d. In each load, 2 pieces out of the 40 are, on average, reworks.

Problem 5. Classify the following activities or events as Required, Appraisal, Prevention, Internal Failure, or External Failure (cost of quality analysis).

a. An acceptance sampling plan examines 200 pieces. If one or fewer is bad, the lot passes. If two or more are bad, the lot requires 100 percent inspection.

b. A lot goes into the above acceptance sampling inspection. Three pieces fail, so the lot receives 100 percent inspection to remove all the bad units.

c. An assembly operation puts two subassemblies together.

d. An SPC chart shows when the process average has shifted.

e. An office chair breaks. The customer identifies the part that needs replacement, but the manufacturer's customer service department does not respond to phone calls or letters. The customer must replace the chair and buys a competitor's product. The customer tells the original supplier he will never buy anything from it again.

f. A transistor assembly process solders the transistor to a stem and places a cap over the transistor.

g. An electrical tester checks the finished units from item f and rejects the bad ones. It also classifies the bad pieces.

h. A product's designers work closely with manufacturing to make sure the product is easy to manufacture. (This is DFM.)

Problem 6. Are the following actions or procedures acceptable under ISO 9000? Why or why not? Should they be changed?

a. A process' operating instruction calls for a 30-minute process time. A manager or engineer verbally instructs a manufacturing shift to set the time to 25 minutes.

b. A process' operating instruction calls for a 30-minute process time. A manager or engineer makes a handwritten change in the operating instruction to set the time to 25 minutes.

c. Operators place bright orange REJECTED stickers on nonconforming pieces and place these pieces on a dedicated shelf. (*Dedicated* means the shelf is used for nothing but rejects.)

d. A factory work area has three shifts, and 12 people do a particular job. An engineer removes the old set of work instructions and replaces them with new ones. The new instructions have the necessary approval from the manufacturing manager. What else, if anything, should have happened?

e. Operators normally use a calibrated electronic micrometer to measure parts. A manual micrometer is sitting on a table in the work area. There is also a window thermometer, but the work area is not subject to temperature and humidity controls.

f. Subassemblies receive bar codes to identify them. The operator scans the bar code at each operation, and a computer logs the subassembly identification and workstation identification. When an operator measures a piece, he or she logs the measurement and scans the bar code.

Solutions
Problem 1

Defect	Abbreviation	Count
Damage	DAM	163
Discoloration	DISC	52
Oxide holes	OX	14
Blisters	BLIS	4

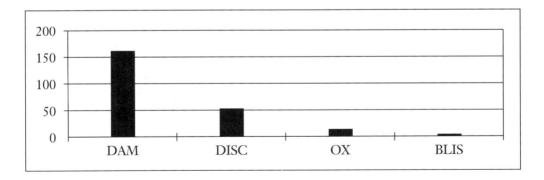

Problem 2

Defect	Abbreviation	Count
Blisters	BLIS	31
Oxide holes	OX	20
Damage	DAM	5
Discoloration	DISC	2
Miscellaneous	MISC	2

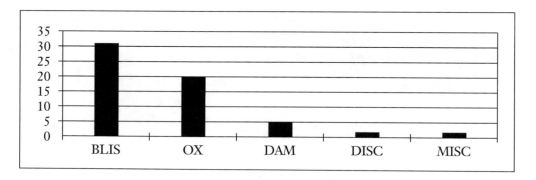

Problem 3

Operation	Abbreviation	Scratches
Silicon nitride etching	NitE	97
Aluminum etching and stripping	Etch	94
Wafer probing (electrical test)	Prob	34
Backside metal deposition	BMet	23
Silicon nitride deposition	NitD	22
Ion implantation	Imp	11
Back grinding	Grind	2
Wafer firing	Fire	0

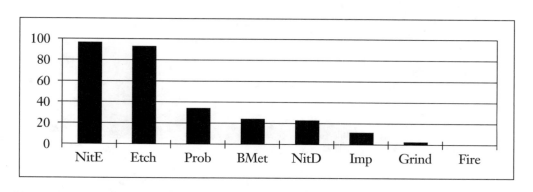

Problem 4

Availability	100%, since the tool is always available.
Operating efficiency	$\dfrac{18 \text{ hours/day (operating)}}{24 \text{ hours/day (availability)}} \times 100\% = 75\%$
Rate efficiency	$\dfrac{40 \text{ pieces/load (actual)}}{50 \text{ pieces/load (capacity)}} \times 100\% = 80\%$
Rate of quality	$\dfrac{38 \text{ good pieces/load}}{40 \text{ pieces/load}} \times 100\% = 95\%$
OEE	$75\% \times 80\% \times 95\% = 57\%$ **or** $$\text{OEE} = \dfrac{\text{Operating time}}{\text{Total time}} \times \dfrac{\text{Good pieces}}{\text{Theoretical output}} \times 100\%$$ $$\dfrac{18 \text{ hours/day}}{24 \text{ hours/day}} \times \dfrac{38 \text{ good pieces}}{50 \text{ pieces/load}} \times 100\% = 57\%$$

Since this operation is not a constraint, don't worry about the idle time (operating efficiency) or partial loads (rate efficiency). Rework, however, is never desirable. The rate of quality goal should be 100 percent.

Problem 5

a. Appraisal; detects quality problems
b. Internal failure; rectification or detailing
c. Required; adds value to the product
d. Prevention; allows adjustment of the process before it makes bad parts
e. External failure Note that the supplier did not suffer an immediate monetary loss (one that would appear in the cost accounting system). Permanent alienation of a customer is an intangible cost with long-term consequences.
f. Required; adds value to the product and is necessary to complete the product
g. Appraisal; separates good pieces from bad ones. The bad pieces are an internal failure cost.
h. Prevention If the product is easy to make, quality problems are less likely.

Problem 6

a. No. If the time should be 25 minutes, the manager or engineer must provide new operating instructions and remove the obsolete ones from the work area. Oral tradition is an unreliable way to communicate work instructions in a factory.

b. No. Handwritten changes are not acceptable, since (1) there is a danger of confusion and (2) they do not guarantee that everyone will be aware of the change. If the process change is urgent, a *written* temporary process change notice is acceptable. At Harris Semiconductor, the engineer, manufacturing leader, and self-directed work team leader sign the temporary change notice. This assures that everyone is aware of the change.

c. Yes. The REJECTED sticker and the segregation procedure assure that no one will use the bad pieces by accident.

d. There needs to be a mechanism for telling the workers about the change and assuring that they have read the new instructions. If they have memorized the old instructions, they may keep doing the job the old way.

e. The manual micrometer is not acceptable and should be removed or placed under calibration control. Otherwise someone might use it to measure parts. The window thermometer is probably acceptable, since the work area is not under temperature and humidity control. Similarly, a wall clock does not need calibration. The key question is, "Could someone use an uncalibrated instrument to measure the work?"

f. Yes. This is an example of product identification and traceability. If there is a problem with the subassemblies, we can identify the workstations that produced them. We can also match measurements (from the gages) with workstations for SPC purposes.

CHAPTER THREE

Statistical Process Control

Business is war; specifically, it is a competition between organizations. Victory in the manufacturing marketplace depends on ability to hit a target or meet specifications. SPC helps us hit the target.

On War ([1831] 1976), a famous book by the Prussian general Carl von Clausewitz, compared war to commerce. It is, he said, "a conflict of human interests and activities." Vince Lombardi, long-time coach of the Green Bay Packers, said that war, football, and business are similar. "The object is to win—to beat the other guy." If several competitors want to sell a product to one customer, there is a conflict or contest. In manufacturing, we win this contest by filling customers' needs. Customers may have needs that their specifications don't cover, but meeting specifications is the first step.

Variation and Accuracy

Two factors define a manufacturing process' ability to meet specifications. These are variation and accuracy. Precision is the opposite of variation.

A manufacturing operation's goal is to make product that meets the customers' needs. Specifications reflect the customers' needs. Product that is in specification is good; we can sell it to the customer. Product that is out of specification is bad; the customer will not accept it. We must rework it or scrap it.

Two factors affect the process' performance. These are variation and accuracy. It is easy to explain these by treating the specification as a target for a gun. Shots that hit the target are good, while those that miss are out of specification. We have a good idea of what accuracy is. If we are shooting at a target, we want to be aiming at its center. Even if we are aiming at the center, however, there is some unavoidable random variation in where the shots hit.

Precision is the opposite of variation. A precise process has very little variation.

A musket is a smoothbore gun that was common in the fifteenth through eighteenth centuries. Soldiers used them in the American Revolution and Napoleonic Wars. A musket has no rifling to make the ball (bullet) spin as it goes down the barrel, so there is a lot of variation in where it hits. It is hard to hit anything with a musket at more than 100 to 150 yards, no matter how good the shooter is. Figure 3.1 shows a simulation of 100 musket shots at a target.

The musket is *accurate*, because its point of aim is exactly in the center of the target. Some shots are off the target, however, because of the tool's *high variation*. We will later see that variation is relative to the target's size or specification width. The musket is not capable of better performance.

We cannot improve the performance of a noncapable tool by adjusting it or looking for a better operator. To improve performance, we must get a better tool or improve the one we have.

How can we improve this process? Three options are (1) Adjust the musket's sights to improve its aim; (2) Hire a better marksman; and (3) Replace the musket with a rifle.

We've already said that the musket's point of aim is the exact center of the target. Any adjustment will make it worse, not better. *A common mistake in industry is to try to improve a process by adjusting it when it does not need adjustment.* Terms for this mistake include *overadjustment* and *tampering*.

Hiring a better marksman won't help either. Putting Daniel Boone or Annie Oakley behind this musket would not improve the results. The computer simulation assumes that a bench rest is holding the musket in place, so

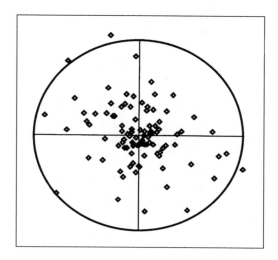

Figure 3.1. Simulation of musket (high variation tool).

there is no operator-induced variation. We also said that the musket is not capable of better performance. *Another common mistake is to blame the tool operator for random variation from a noncapable tool.*

A rifle barrel causes the bullet to spin as it goes down the barrel. This makes the bullet go in a straight line. When a quarterback throws a good pass, the football spins and goes in a straight line. If a defender hits the quarterback as he throws the pass or forces him to hurry, the football may tumble end over end. It usually doesn't go where the quarterback wants it to go. A rifle bullet in flight acts like a good football pass, and a musket ball acts like a bad one. A spinning projectile has much less variation than a tumbling one. While a musket's extreme range is 150 yards or so, a modern rifle is effective at up to 750 or 1000 yards. Figure 3.2 simulates a rifle and uses the same target that the musket used.

The terms *capable* and *not capable* have special meanings in industrial statistics. A tool's *process capability index* measures its ability to meet specifications. The capability index uses the ratio of the specification width to the variation.

$$\text{Process capability index } = \frac{\text{Specification width}}{\text{Variation}}$$

We will not get into the specific mathematical formulas here, but Table 3.1 classifies the results. "Bad pieces per million" assumes that the process is on target, or halfway between the specification limits.

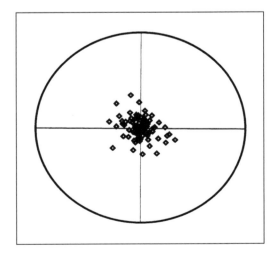

Figure 3.2. Simulation of rifle (low variation tool).

Table 3.1. Process capability indices.

Process capability index	Process status	Bad pieces per million	Similar to a(n)
Less than 1.00	Not capable	2700 or more	Musket
1.00 to 1.33	Marginal	2700 to 63.3	
1.33 to 2.00	Capable	63.3 to 0.002 (2 per billion)	Rifle
2.00 or better	Extremely capable	0.002 or less	Olympic match rifle

Process Capability and Tight Specifications

Schoolchildren learn a simple explanation of how Americans won the Revolutionary War. The smart Yankees dressed in brown clothing and hid behind trees. The unfortunate British soldiers wore bright red coats and stood in the open to fight. The British uniform included a white crossband across the chest, which made a superb target. The gold plate on the front of the soldier's mitre hat was another. In case their opponents didn't see them coming, they beat drums that were audible for miles. Scottish regiments added bagpipes, which may have terrified some adversaries, but did not promote secrecy of movement.

The problem with this stereotype is that the redcoated soldiers ended up owning half the world. Wellington's Redcoats handily whipped Napoleon. This suggests that the British were actually very smart, but their methods didn't work in North America. The British were unsuccessful in North America because of a process capability issue.

The British smoothbore musket, the Tower musket, could fire five or six shots a minute. The secret was a loose-fitting musket ball, which was easy to ram down the barrel. The loose fit, however, made this weapon's variation even worse. Nonetheless, it worked well in Europe, where troops lined up in shoulder-to-shoulder formations. An enemy regiment was a big target; that is, the specification was wide. An English regiment could fire 10,000 shots in the enemy's general direction every minute. It was almost futile to aim, but the enemy's shoulder-to-shoulder formation gave each shot a chance to hit. Ten thousand chances a minute added up to a lot of damage.

In Figure 3.3, 10 Redcoats fire 50 shots at the French in one minute. (This was usually whom the English were fighting in the eighteenth century.) They are aiming at the man in the center* and score multiple hits. Meanwhile, many

*In reality, the soldiers would not all aim at one opponent. The idea of the simulation is to show the soldier's chance of hitting the opponent he is aiming at. Even if he misses, the shot may hit the target's neighbor.

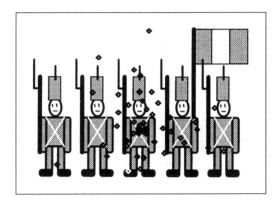

Figure 3.3. High variation process versus wide specification.

of the shots hit his neighbors. Against a big target, or wide specification, this method is very effective ("If you can't shoot well, shoot a lot").

Muzzle-loading rifles were available, and these were effective at 300 or more yards. They were popular among hunters, whom armies often recruited for rifle regiments. The French and Germans called these troops Chasseurs and Jaegers, meaning "hunters." Muzzle-loading rifles could not fire quickly, since they required a tight fit between the bullet and barrel. It took a long time to ram the bullet down, and this made rifle-armed infantry very vulnerable on an open battlefield. Their initial volley would cause a lot of damage, but enemy infantry with muskets could fire several times before the riflemen could reload. Alternately, a bayonet charge or cavalry assault could overrun them before they could fire a second time. Therefore, armies used riflemen in forests and rough ground where it was hard for regular infantry or cavalry to move. The populated areas of Europe did not have dense forests, so rifle troops did not play a large role.

In the simulation in Figure 3.3, 10 riflemen would probably have scored 10 hits in one minute. The musket troops, however, hit their intended target more than 10 times (plus several hits on his neighbors) despite the musket's high variation. Rapidity of fire was more important than precision.

North America had very dense forests, and the inhabitants were skilled foresters and sharpshooters. The typical French or English colonist had a rifle for hunting and for protection against hostile Indians. Colonists learned woodcraft from friendly American Indians, who were excellent foresters. The dense forests made cavalry almost useless and made it hard to move cannon and supply wagons. An American in a brown jacket and coonskin cap behind a tree was hard to see, let alone hit. The British had to meet a tight specification, but their process (musket fire) could not do it.

In Figure 3.4, 10 Redcoats fire 50 shots in one minute at a Revolutionary who has exposed part of his head to aim his rifle. This is a very small target (tight specification), and the British score perhaps six hits. Even this performance requires the optimistic assumption that the British soldiers can aim at their opponents. The black powder smoke from both sides' weapons, and the colonists' refusal to fight in the open while wearing brightly colored uniforms, made aiming problematic at best.

Meanwhile, 10 opposing riflemen would probably have scored 10 hits on the British. In Figure 3.5, the riflemen score 9 or 10 hits. Again, they would

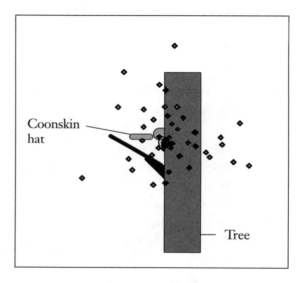

Figure 3.4. High variation process versus tight specification.

Figure 3.5. Low variation process versus wide specification.

not actually aim at only one opponent. Centering the shots on one figure shows the tight spread of the rifle shots.

New industrial products often have tighter specifications than old ones. This is especially true in the electronics industry. Fifteen years ago, a computer chip wire might be 6 microns* wide. Today, half-micron line widths are common. Suppose the specification is ±5 percent. If the target is 6 microns, the specification is [5.70, 6.30] microns ("5.70 to 6.30 microns"). The specification is 0.60 microns wide. If the target is 0.5 microns, the specification is [0.475, 0.525] microns. This specification is 0.05 microns wide. Figure 3.6 shows that *the smaller target demands a more capable tool.*

The picture on top is like a shoulder-to-shoulder formation, and the one on the bottom is like someone hiding behind a tree. A tool that can handle the [5.70, 6.30] specification may not be able to handle the [0.475, 0.525] specification.

Quantity Versus (?) Quality

The simulations in the previous section involved 50 musket shots (high quantity, low quality) versus 10 rifle shots (high quality, low quantity) in one minute. Is there always a quantity/quality trade-off, as Figure 3.7 suggests?

During the Revolutionary War, a Scottish captain, Patrick Ferguson, invented a breechloading rifle. The Ferguson rifle could fire six or more shots a minute, and it had the range of a Kentucky or Pennsylvania rifle. This provided a third alternative. Figure 3.8 simulates 60 shots from 10 soldiers

---

*A micron is a thousandth of a millimeter, and a millimeter is the smallest division on a metric ruler.

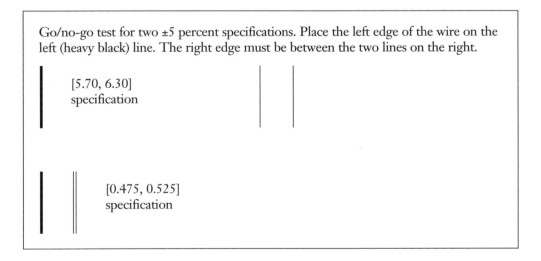

Figure 3.6. Shrinking wire sizes in microelectronics manufacturing.

using Ferguson rifles. Fortunately for the American Revolution, the British Army ignored the invention. Ferguson used his own money to equip a company of riflemen with it, and these troops inflicted horrific damage on the Americans. When Ferguson died in a battle, however, his invention died with him. It took another 60 or 70 years to reinvent the breechloading rifle.

The goal of modern industrial equipment is to produce quantity and quality. It is not an either/or trade-off.

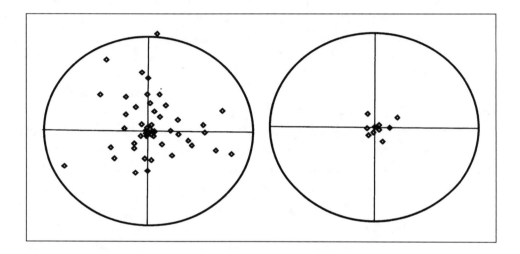

Figure 3.7. Quantity or quality?

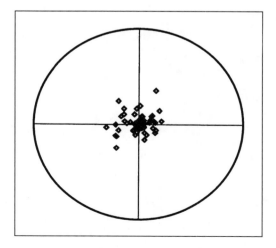

Figure 3.8. Quantity and quality.

Accuracy

An accurate process aims at the center of the specification. The goal of target shooting is to hit the bull's-eye. In manufacturing, we want to hit the center of the specification. The *nominal* measurement or dimension is halfway between the upper and lower specification limits.

Consider a rifle whose sights are out of adjustment. The aiming point will be somewhere other than the bull's-eye. Figure 3.9 shows this.

The rifle is precise (because of its low variation), but it is not accurate. In contrast, the musket was in control because its aiming point was the bull's-eye, but it was not capable. Figure 3.10 compares capability and control.

Consider the rifle in Figure 3.9. A target shooter would correct this situation by lowering the rifle's back sight.* We can also adjust manufacturing equipment to bring the process back on target. This might involve changing the process time, adjusting a gas flow, or replenishing a chemical. SPC tells us when to adjust the process.

A target shooter would not fire 100 shots before adjusting the sights. The first five shots show that we should probably move the rear sight down (see Figure 3.11).

The group of five rifle shots is a *sample* of the rifle's performance. This sample tells the shooter whether to adjust the process by moving the sight.

---

*A rifle's front sight is stationary. The shooter adjusts the aiming point by raising or lowering the back sight, or moving it right or left.

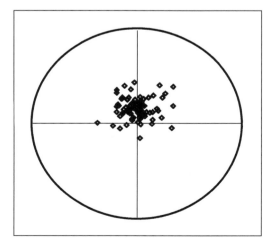

Figure 3.9. Process shift; misaligned rifle sights.

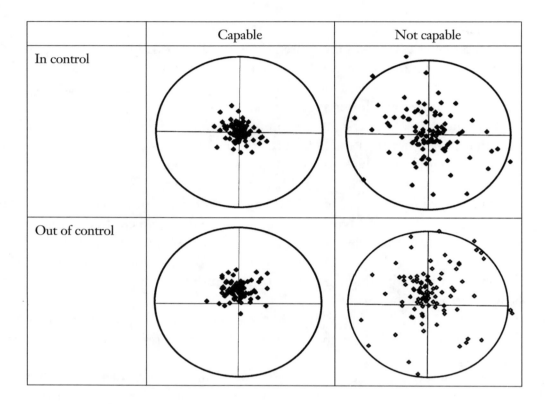

Figure 3.10. Capability and control.

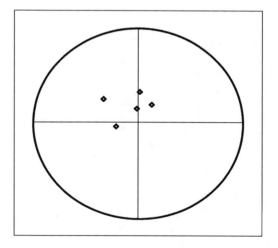

Figure 3.11. Group of five rifle shots; rear sight too high.

SPC involves samples of the process' performance. These samples tell us whether to adjust the manufacturing process.

Overadjustment or Tampering

Don't adjust the process when it's working properly. ("If it ain't broke, don't fix it.")

We might ask, "Why fire five shots before adjusting the rifle sights? Wouldn't adjusting them after every shot make the process work even better?" Let's try this. The computer simulation adjusts the aiming point after every shot. For example, if the shot hits one inch high and one inch right of center, the program lowers the aiming point one inch and moves it left an inch. Figure 3.12 shows what happens.

This is clearly much worse. The adjustment procedure is actually trying to compensate for random variation. It tries to "fix" the process when there is nothing wrong with it.

An *iatrogenic disease* is a disease that the doctor causes. *Iatros* is Greek for *physician,* and *genic* means *caused by.* A doctor can cause an iatrogenic disease by prescribing medicine that the patient doesn't need. At best, the prescription wastes money, and it can cause undesirable side effects. At best, adjusting a process that is running properly wastes manufacturing personnel's time. The simulation shows that it also can induce excessive variation—the manufacturing equivalent of an iatrogenic disease.

Random Variation and Assignable Causes

How do we know when we should fix the process? Control charts, which we will look at later, distinguish random variation from assignable causes or special causes.

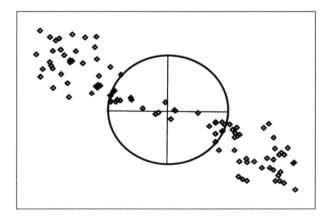

Figure 3.12. Overadjustment or tampering.

Manufacturing problems (rework, scrap, and defects) come from two sources. These are *common causes* or *random variation* and *assignable causes* or *special causes*. Common causes are inherent in the process. The musket's extreme variation is an example. The only way to reduce or get rid of random variation is to improve the process. An assignable or special cause is something that is wrong with the process. The misaligned rifle sights are an example.

To reduce or get rid of common causes, we must improve the process. To remove an assignable cause, we must fix the process. We should always look for ways to improve the process. SPC tells us whether we need to fix the process.

This shows that the statement "if it ain't broke, don't fix it" does not conflict with the idea of continuous improvement. There is a big difference between improving something and fixing it. Improving it means changing or replacing the process to make it work better. The process or system that we change or replace may have worked properly, or as expected, but something better is available. In contrast, fixing something means correcting something that was wrong with it. Table 3.2 shows examples. The correction examples all refer to corrections of assignable causes or special causes.

The Normal Distribution

Measurements from manufacturing processes usually follow the normal (bell curve) distribution. This describes the chance that a part from the process will have dimension or measurement x. The variable x is a real number, and we want it to be within the specifications.

Table 3.2. Improvement versus correction.

Examples of improvement	Examples of correction (fixing)
Evolution of ground transportation • Horse and buggy • Early automobile (Model T) • Ford Taurus, Chevrolet Caprice, Chrysler Concorde • Future: George Jetson's car?	• Call veterinarian to cure sick horse • Take car to garage to fix leaking radiator
Evolution of electronic technology • Vacuum tube • Transistor • Integrated circuit	Replace burned-out vacuum tube.
Musket → Rifle	Adjust misaligned rifle sights.

There is a relationship between the normal distribution and the shot patterns on the targets. The distribution's mean, or center of gravity, is the aiming point. We want it on the nominal or bull's-eye. The distribution's variation is like the spread of the shots. We want the distribution to act like a rifle and not a musket.

Consider a semiconductor manufacturing operation in which we want to deposit a 1000 Å film of silicon dioxide on a silicon wafer. Because of variation, we know that not all the wafers will get exactly 1000 Å of silicon dioxide. If we process 500 wafers, how many will have a 970 to 980 Å film? How many will have 1020 Å or more? If the specification is [950, 1050], how many will be out of specification? The normal distribution (bell curve) allows us to predict this if we know the two parameters in Table 3.3. (A *parameter* is a number that describes the distribution. Don't worry about the Greek letters; there's no need to remember them.)

Figure 3.13 is a simulation of the silicon dioxide process. The mean is 1000, and the variation is very large ($\sigma = 20$). The height of each bar shows the number of wafers with each measurement range. Suppose that each bar represents a stack of wafers and that each wafer has the same weight. Imagine that the axis is a weightless balance beam. If we place a fulcrum at the *center of gravity*, or *mean*, the beam will balance. We will need another simulation to show the significance of the variation.

Note that a plot of the expected number of wafers in each measurement range looks like a bell. This is the normal distribution.

The mean, or aiming point, is the 1000 Å nominal (bull's-eye). There is, however, a lot of spread in the wafer measurements. This process is actually the manufacturing equivalent of a musket. Figure 3.14 shows the correspondence between the histogram and the shot pattern on a target. It also shows

Table 3.3. Parameters of the normal distribution (bell curve).

Parameter	Significance
Mean (μ or mu; Greek *m* for mean)	The process' center of gravity or aiming point. It is similar to the average. Technically, the *mean* refers to the entire process, and the *average* refers to a sample from the process.
Variation (σ or sigma; Greek *s* for standard deviation)	The *variance* is the square of the standard deviation (σ^2). This term is easy to remember because it sounds like "*variation.*" All you need to remember is that a large standard deviation, or a large variance, means a lot of variation. This is undesirable in manufacturing.

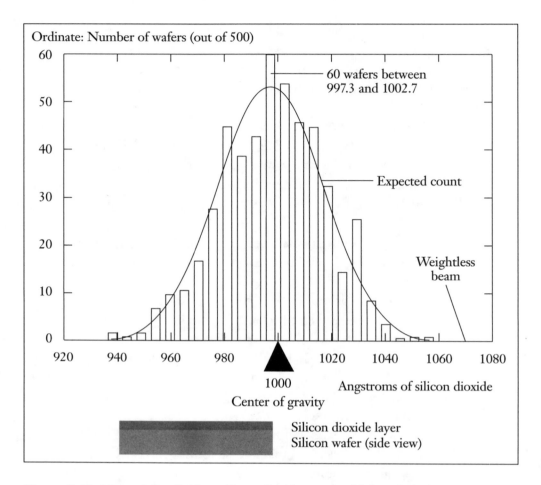

Figure 3.13. Normal distribution, silicon dioxide process (high variation).

the effect of variation. In manufacturing, the target is usually one-dimensional. We've used circular targets because they help explain variation and accuracy. The histogram shows in one dimension what the target shows in two dimensions. The nominal is the bull's-eye, and the histogram's spread is like the shot pattern's spread.

Suppose that the process mean shifts from 1000 to 1010 Å. Figure 3.15 shows the histogram for 500 wafers. The corresponding target shows 100 shots from a rifle. This process is capable because its variation is low, but it is out of control.

There are SPC charts that correspond to these histograms and targets. The charts can tell us whether a musket has replaced our rifle. That is, has the process variation increased? SPC charts also tell us whether the process needs adjustment. Has the process mean increased or decreased? If so, we can adjust the process, just as we can adjust the sights on a rifle.

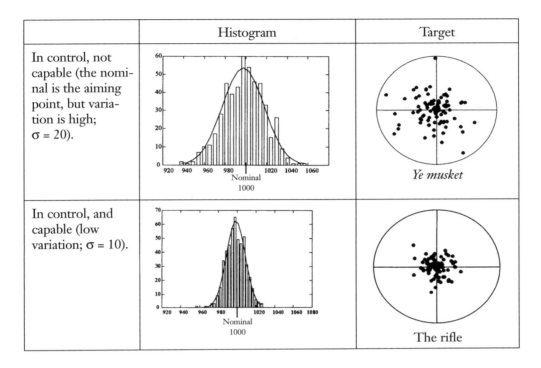

Figure 3.14. Histograms and targets.

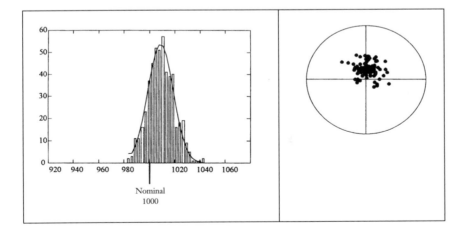

Figure 3.15. Shift in process mean (capable, out of control).

The Normality Assumption

Traditional SPC relies on the assumption that the sample data follow the normal distribution. All statistical tests have risks for false alarms and for not giving alarms when they should. The risks for the traditional (Shewhart) control chart assume that the distribution is normal. If the data don't follow a normal distribution, we must use alternative procedures.

The procedures for handling non-normal data are beyond the scope of this book. Readers should, however, be aware of this consideration. If a histogram of the process data doesn't look like a bell curve, the traditional methods won't work properly. An engineer or industrial statistician needs to look at the data.

A one-sided specification is often a warning that the population will not follow the normal distribution. For example, an impurity level in a chemical will have an upper specification. The customer won't accept more than 10 ppm impurity, but will be very happy with zero. It isn't possible to have less than zero impurities. Figure 3.16 simulates 100 chemical batches with an average impurity level of 4 ppm.

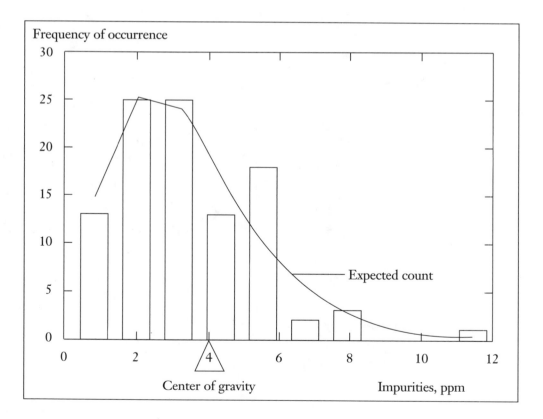

Figure 3.16. Non-normal distribution (mean = 4, σ = 2).

Feedback Process Control

Feedback process control means (1) observing data from the process, (2) adjusting the process accordingly, and (3) observing data again to make sure that the adjustment was effective. SPC is a form of feedback process control. The idea of feedback control is at least 350 years old. Miyamoto Musashi, a Japanese samurai warrior, identified the principle in *A Book of Five Rings* ([1645] 1974). English translations of this book were popular in the early 1980s, because the Japanese were using it as a business management guide.

In feudal Japan, the bow was the traditional missile weapon. Portuguese traders introduced the Japanese to the arquebus (matchlock musket) in the sixteenth century. Musashi wrote that the bow was superior to the musket, because the archer can watch his arrows in flight and change his aim accordingly. The arrow gave the archer feedback on wind conditions. The information might even let him correct his aim and shoot a second arrow before the first one landed. This is like using tracer bullets to aim a modern firearm.

The arquebusier, or matchlock gunner, cannot see the bullets in flight. In a battle, he cannot even tell whether his shots are hitting the enemy. If he fires alongside other gunners, no one can tell whose shots hit and whose miss. This lack of data traceability, which we examined briefly in chapter 2, makes it even harder for gunners to perform effectively (Figure 3.17). They can replace or rework their nonconforming (off-target) shots by reloading and firing again. This is not, however, something one wants to do when the enemy is returning the fire or advancing with swords and lances. We don't want to do it in manufacturing either, when a competitor is after our market share.

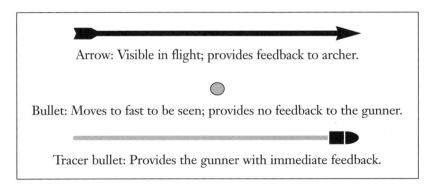

Arrow: Visible in flight; provides feedback to archer.

Bullet: Moves to fast to be seen; provides no feedback to the gunner.

Tracer bullet: Provides the gunner with immediate feedback.

Figure 3.17. Feedback.

The musket was actually inferior to the bow and superseded it only because it was a low-skill weapon. It took a lifetime of training to make an archer, while a few lessons would make someone a passable musketeer. This is why the warrior caste (samurai) objected to firearms. The samurai spent their lives learning military skills and resented a weapon that required little skill. Only a samurai could use a sword or bow well enough to kill another samurai. The samurai's training, however, could not stop a bullet, and anyone with a musket could fire one. Anyone—a peasant, artisan, merchant, or even an eta (untouchable or outcast)—could pull a trigger.

Figure 3.18 shows a feedback process control loop. The key aspects are

- Measuring the process' output
- Comparing the measurement to the desired conditions
- Controlling action, if necessary, to achieve the desired conditions

In the archery example, the archer

- Watches (measures) his arrows in flight.
- Compares them against the desired condition (heading toward target).
- Adjusts (controls) his aim accordingly.

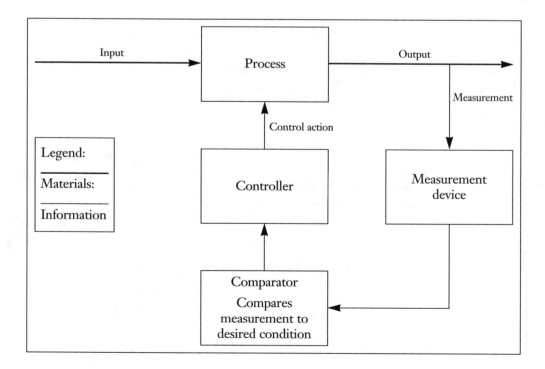

Figure 3.18. Feedback process control.

A household thermostat is another example of a feedback process control loop. Figure 3.19 shows how the system

- Measures the temperature.
- Compares the temperature against the desired condition (set point).
- Adjusts the condition by turning the furnace or air-conditioning on or off.

The word *loop* is important. The process control loop acts to correct any difference between the desired and actual conditions. *Closing the loop* means acting and measuring again to confirm the effectiveness of the action. Measurement has little value without action. If a household thermostat reads 50°F (10°C), the information might be interesting, but the people in the house would still be cold. Action without measurement can be worse than useless. If the furnace acted without information from the thermometer, the temperature might rise to 95°F (35°C) and drive the people out of the house. Measurement and action together make up an effective process control loop.

In continuous processes, like most chemical manufacturing processes, control loops are automatic. Continuous processes, however, act on products

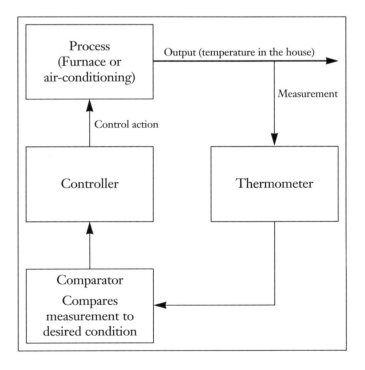

Figure 3.19. Household thermostat.

that flow or pour. These are usually liquids or gases. Manufacturing processes that act on discrete workpieces, or "widgets," require SPC.

Figure 3.20 shows the SPC control loop. Control action often requires a manual process adjustment. For example, an operator might adjust the process time to change the silicon dioxide thickness. In a chemical etching or plating operation, the operator might replace the chemical. In a machine shop, the operator might replace a worn drill bit or die.

The process or operating instructions should include directions for routine process adjustments. Harris calls these directions the *out-of-control action procedure*, or OCAP. Hradesky (1988) calls them the *corrective and preventive action matrix*, or CP matrix. The idea is to allow the operator to adjust the process and continue operations without having to consult a technician or engineer.

As the manufacturing team (operators, technicians, and engineers) identifies and solves new problems, it should add these to the OCAP or CP matrix. If the problem recurs, the operator can then fix it immediately. This is like developing immunity to certain diseases. When we get one of these diseases, we feel sick while our bodies fight it. Afterward, our immune systems remember the bacteria or virus that caused the problem and how to destroy

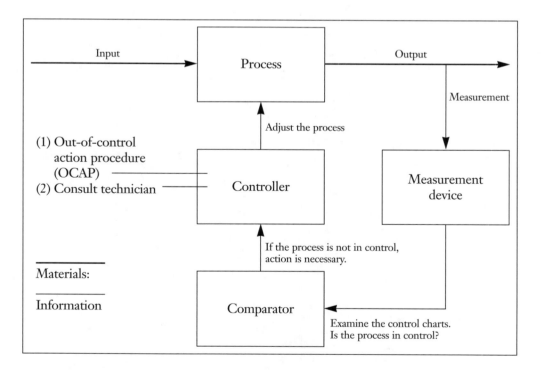

Figure 3.20. SPC control loop.

them. If the disease shows up again, our bodies destroy it immediately and we never notice it. The OCAP or CP matrix does the same for a manufacturing process. If the manufacturing team anticipates possible problems and develops procedures for handling them, this is like vaccination. The process never gets sick in the first place.

Requirements for Successful SPC

There are four requirements for effectively using SPC. These are

1. Data integrity
2. Data traceability
3. Identification of critical process parameters
4. Real-time capability

They are defined in Table 3.4.

Figure 3.21 shows the idea of data traceability. Suppose that the measurement from the gage shows a problem with the process. This information is most useful if we can identify the workstation that is causing the problem.

Traceability requires an identification system for the lots or pieces. There also must be a system for recording where the pieces have been. In semiconductor manufacturing, wafers often have serial numbers for identification. Product lots may have serial numbers, bar codes, or other identification. Which set of information in Table 3.5 is more useful?

Table 3.4. Requirements for SPC (Messina 1987, 1–2).

(1) Data integrity	Data (measurements) must be accurate. We will later examine *gage capability*, which is the ability of gages or instruments to make accurate and repeatable measurements.
(2) Data traceability	This means being able to trace measurements to the processes, equipment, and material that produced them.
(3) Identify critical process parameters	Identify the process steps that have significant effects on product quality. Harris Semiconductor calls these *critical nodes*.
(4) Real-time capability	Feedback must be prompt enough to allow timely process adjustments. An arrow, or a stream of tracer bullets, provides immediate feedback to the archer or gunner. We want similar real-time feedback for a manufacturing process.

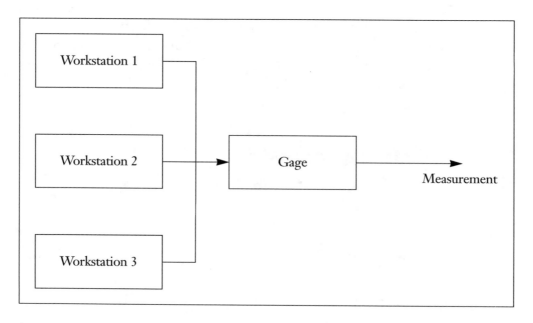

Figure 3.21. Data traceability.

Table 3.5. Nontraceable and traceable data.

Without traceability			With traceability		
Wafer	Oxide thickness		Wafer	Oxide thickness	Oxide workstation
8470021	975 Å		8470021	975 Å	2
8470350	998 Å		8470350	998 Å	2
8470226	1002 Å		8470226	1002 Å	3
8470148	982 Å		8470148	982 Å	2
8470223	1010 Å		8470223	1010 Å	1
8470104	1005 Å		8470104	1005 Å	3

Figure 3.22 shows the idea of real-time feedback. To be useful, process measurements must be timely. Suppose that the final electrical test usually happens about a week after operation 1. The test provides information about what operation 1 has done. If it identifies a problem, however, the operation will have had a week to run without correction. The information arrives too late to do much good. If the intermediate measurement happens right after operation 1, it provides immediate feedback. *To control a critical operation, place the measurement as close to the operation as possible.*

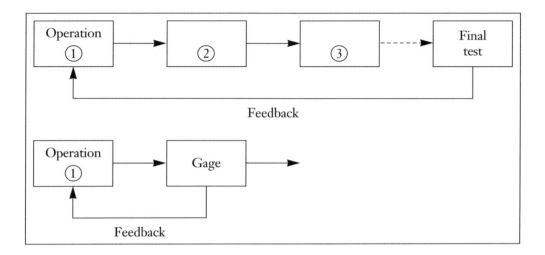

Figure 3.22. Real-time feedback: Which control loop would you rather have?

Statistics

Control charts use three statistics.

1. The average reflects the process' mean or center of gravity.
2. The range reflects the process' variation.
3. The standard deviation also reflects the process' variation.

Consider a sample of n measurements. Here, $n = 5$.

$$9.1 \quad 10.5 \quad 10.9 \quad 9.8 \quad 10.5$$

Average

The *average* is the sample's arithmetic average or center of gravity. It reflects the process mean. The average is simply the arithmetic average of the measurements. The symbol for a sample average is $\bar{x}$ ("x bar"). The horizontal line, or bar, is the statistical symbol for average. The SPC chart that shows the sample averages is an *x bar chart* or $\bar{x}$ chart. (If the samples have only one measurement each, the chart is an *X chart*; *X* refers to an individual measurement.)

$$\bar{x} = \frac{9.1 + 10.5 + 10.9 + 9.8 + 10.5}{5} = \frac{50.8}{5} = 10.16$$

The average is the sample's center of gravity. Figure 3.23 treats each number as a weight. Place each weight at the number's position on a weightless number line. Place a fulcrum under the average, and the number line or beam will balance.

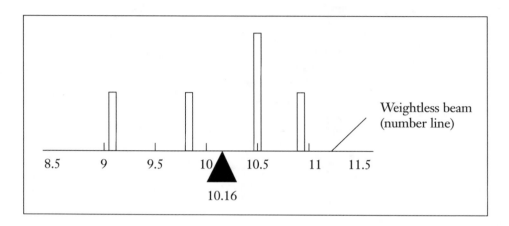

Figure 3.23. Sample average as center of gravity.

A target also shows this principle in two dimensions. To sight in a rifle or pistol, the shooter actually estimates the shot group's center of gravity. Figure 3.24 shows the idea. If the group is significantly off center, the shooter adjusts the sights to bring it on center.

Range

The *range* (R) is the difference between the largest and smallest measurements in the sample. It reflects the process' variation.

The target in Figure 3.24 also shows the distance between the two shots that are furthest apart. It is reasonable to expect this distance to depend on the process' variation. If the variation is large, we expect the distance between the shots to be large. If the variation is small, we expect the distance to be short. Range, or distance between the largest and smallest numbers, measures variation for one-dimensional systems. The two-dimensional target only illustrates the concept. A two-dimensional system does, however, have a center of gravity.

In the sample of five measurements, 10.9 is the largest and 9.1 is the smallest. The range is 10.9 − 9.1 = 1.8. Figure 3.25 shows that the range is simply the distance between the largest and smallest measurements. The SPC chart that shows the sample ranges is an *R chart*. The five measurements are 9.1, 10.5, 10.9, 9.8, and 10.5.

Standard Deviation

The *standard deviation* (s) also reflects the process' variation. The range is easy to calculate, but it uses only two of the measurements. It sacrifices the information that is available from the other measurements in the sample.

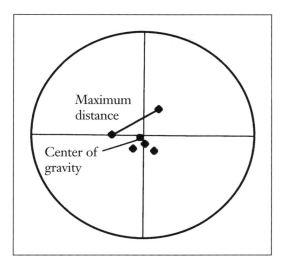

Figure 3.24. Shots on target; center of gravity and range.

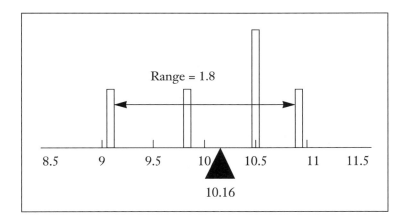

Figure 3.25. Range in a sample.

The standard deviation uses all the measurements to estimate the process variation. To understand the concept, look at the target in Figure 3.26. The idea is to use the distances between the individual shots and their center of gravity. If there is a lot of variation, the sum of these distances will be large. If there is little variation, the sum of these distances will be small.

Now consider the numbers in the sample. The standard deviation uses the distances in Figure 3.27. Since two measurements were 10.5, the calculation uses the distance between 10.5 and 10.16 twice.

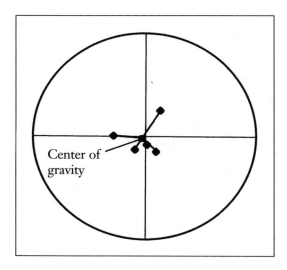

Figure 3.26. Concept behind the standard deviation.

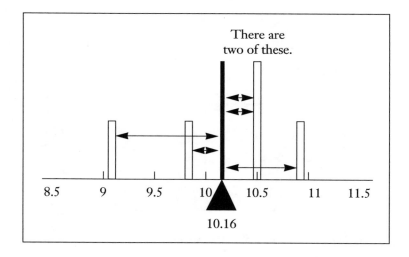

Figure 3.27. Concept behind the standard deviation; one dimension.

The standard deviation's practical drawback is that it takes extra time to calculate. It is not convenient for hand calculations, and it is even inconvenient to do routinely on a calculator. This is why most noncomputerized control charts use the range instead of the standard deviation.

Here is the formula for the sample's standard deviation. You don't have to remember it or use it. A control chart that uses standard deviation will probably be on a computer, which calculates it automatically. The symbol for the sample standard deviation is "*s*," and the SPC chart that uses it is an *s chart*. The *s* chart and *R* chart do not appear together, because they do the same job. They monitor the process variation.

Formula for Sample Standard Deviation
This formula is given for information purposes only.

$s = \sqrt{\dfrac{1}{n-1}\Sigma_{i=1}^{n}(x_i - \bar{x})^2}$	Σ is Sigma (capital Greek S), and it means "to sum." For each measurement in the sample of *n* measurements ("*i* = 1 to *n*," where x_i is the *i*th measurement), add the squares of the differences (distances) between the individual measurements and their average. Then divide the sum by $n - 1$.

In the example,

$$s = \sqrt{\frac{1}{5-1}\left(\begin{array}{l} (9.1 - 10.16)^2 + (10.5 - 10.16)^2 + (10.9 - 10.16)^2 + \\ (9.8 - 10.16)^2 + (10.5 - 10.16)^2 \end{array} \right)}$$

$$= \sqrt{\frac{1}{4}((-1.06)^2 + 0.34^2 + 0.74^2 + (-0.36)^2 + 0.34^2)}$$

$$= \sqrt{\frac{2.032}{4}} = 0.713$$

Again, workers are unlikely to have to do this on a manufacturing line. Hand calculators are available that will calculate *s* automatically. All the user needs to do is type in the measurements. Readers should remember that both the *s* chart and *R* chart monitor the process variation.

Outliers
Outliers are unusual measurements that require investigation. Either the measurement is incorrect or something is wrong with the manufacturing process. Outliers are unusual measurements that are out of line with the rest of the data. Consider Figure 3.28.

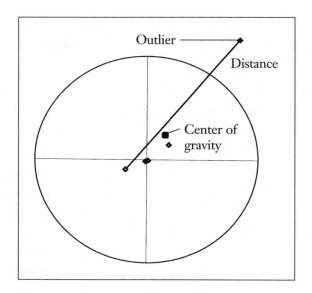

Figure 3.28. Outlier.

The point outside the circle isn't an outlier because it's outside the circle. We've seen that a process with high variation may place several points outside the target. These points come from the same process that produced the ones inside the target. It's an outlier because it is far from the other four points. It's unlikely that this point came from the same process that produced the other four. Therefore, something unusual happened.

The outlier greatly distorts the average and the measurements of variation. The range between the two most distant points is very large. The outlier also inflates the standard deviation. Figure 3.29 shows the same set of points without the outlier.

Figure 3.30 shows a one-dimensional example. The measurements are as follows:

$$9.4 \quad 14.0 \quad 10.3 \quad 9.8 \quad 10.3$$

$$\bar{x} = \frac{9.4 + 14.0 + 10.3 + 9.8 + 10.3}{5} = \frac{53.8}{5} = 10.76 \text{ and } R = 14.0 - 9.4 = 4.6$$

Without the 14.0,

$$\bar{x} = \frac{9.4 + 10.3 + 9.8 + 10.3}{4} = \frac{40.2}{4} = 9.95 \text{ and } R = 10.3 - 9.4 = 0.9$$

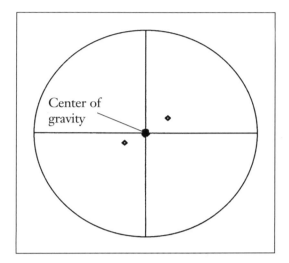

Figure 3.29. Same target without outlier.

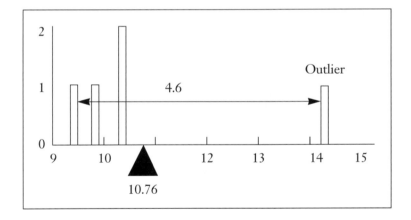

Figure 3.30. Data set with outlier.

The outlier exerts a lot of *leverage* on the sample statistics. This shifts the average, or center of gravity, to the right. The range is also very large. Range is especially sensitive to outliers, since it uses the largest and smallest measurements.

Outliers are unusual measurements that require investigation. Consider remeasuring the specimen. If the measurement does not change, investigate the process that produced the specimen.

Control Charts

Control charts tell us whether the manufacturing process needs adjustment. A control chart is a graphical method for assuring the consistency of a process.

Risks

Control charts are statistical tests in a graphic (picture) form. Whenever we use a statistical test, there are two risks. There is a risk of a false alarm and a risk of not detecting a problem when there really is one. Table 3.6 describes the risks associated with statistical tests.

A test's power improves with sample size. Tests also become more powerful as the situation gets worse. As the wolf gets closer, the shepherd is more likely to see it.

Table 3.6. Risks associated with statistical tests.

State of nature (actual situation)	Decide that there is a problem	Decide that there is no problem
There isn't a problem; the situation is as it should be.	**False alarm risk** • The risk of crying wolf when there isn't one • Risk of convicting an innocent defendant • Quality acceptance sampling: risk of rejecting a good lot • SPC: risk of calling the process out of control when it is in control	**100% – false alarm risk** • Chance of acquitting an innocent defendant • Quality acceptance sampling: chance of accepting a good lot • SPC: chance of calling the process in control when it is
There is a problem; the situation requires adjustment.	A test's ability to detect a real problem, or difference, is its power. • Chance of seeing the wolf coming • Chance of convicting a guilty defendant • Quality acceptance sampling: chance of rejecting a bad lot • SPC: chance of calling the process out of control when it is	**Risk of missing the problem** • Risk of not seeing the wolf • Risk of acquitting a guilty defendant • Quality acceptance sampling: risk of shipping a bad lot • SPC: risk of calling the process in control when it is out of control

Here is an example that involves quality acceptance sampling. The goal is to take a sample from a large lot and accept lots that have 1 percent or fewer bad items. We want to reject lots that have 3 percent or more bad units. One percent is the acceptable quality level (AQL); 3 percent is the rejectable or unacceptable quality level, or lot tolerance percent defective (LTPD). Readers with experience in acceptance sampling will recognize terms; however, it is not necessary to remember them.

Now consider three sampling plans: (n = 50, c = 1), (n = 100, c = 2), and (n = 200, c = 4). The sample size is n, and the acceptance number is c. For example, n = 100, c = 2 means to inspect 100 pieces and accept the lot if two or fewer are bad. Reject the lot if three or more are bad. In each plan, the acceptance number is 2 percent of the sample size. Which plan is best at distinguishing an acceptable lot from an unacceptable one? Figure 3.31 shows the powers of the three sampling plans versus the condition of the lot. (Experienced quality inspectors will recognize the curves as reverse operating characteristic curves, or OC curves. A sampling plan's OC curve shows the chance of accepting the lot as a function of its condition.)

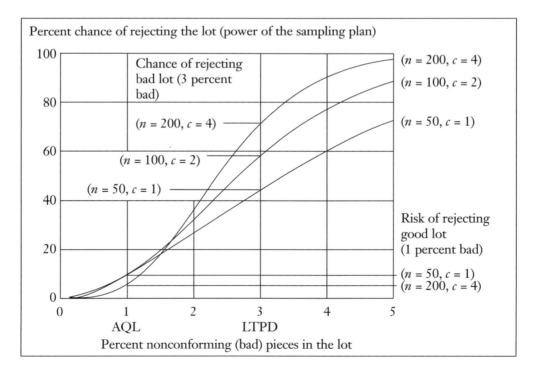

Figure 3.31. Powers of sampling plans.

Observe that of the three plans, the ($n = 200$, $c = 4$) plan is least likely to reject a good lot and most likely to reject a bad one. Plans with large samples are better at telling good lots from bad ones. In SPC, plans with large samples can better detect processes with assignable or special causes.

Also note that as the lot becomes worse (the wolf gets closer), all the plans become more likely to reject it. It is easier to detect a big problem than a little one. This also applies to SPC.

Taking the Sample

Data collection is the keystone of SPC. No statistical control method can be better than the information that goes into it.

How to collect data for SPC

1. The process' operating instruction should include the sampling method. The instructions will say how many pieces to measure and what to measure.

2. Perform the measurements carefully. If the numbers don't look right, don't assume they are wrong.

   a. Unusual measurements may be outliers.

   b. Try repeating the measurement. If the same number is obtained, there is an outlier.

   c. Outliers are evidence of assignable or special causes in the process or of problems with the gage.

      1. If one specimen out of several produces an outlier, but the other numbers are typical for the process, something unusual happened to the specimen. This means an assignable cause in the process.

      2. If all the measurements in the sample are outliers (unusual for the process), the gage could be at fault.

3. Record the results accurately.

   a. Suppose a gage measures samples from several processes or workstations. Be sure to assign the measurement to the process and workstation that produced it.

      1. If the processes' specifications differ widely, mistakes should be obvious. A process whose nominal is 300 Å will not produce a 490 Å result, unless something is seriously wrong with it. There may be, however, a process whose nominal is 500 Å.

This is probably where the 490 Å measurement came from. A 490 Å measurement from a 300 Å nominal process is an outlier and deserves a second look.

2. If the processes (or workstations) are trying to meet similar specifications, mistakes are not obvious. Special care is necessary to make sure the entries are correct.

b. Double-check the numbers, especially when typing them into a computer.

1. Forgetting (or adding) a zero will be obvious. For example, entering 100 instead of 1000 will produce an outlier.

2. Pressing the wrong key for the first digit may produce an obvious error. For example, entering 650 instead of 550 is likely to produce an out-of-control warning. Investigation will quickly reveal the error.

3. Pressing the wrong key for the second or third digit may not produce an obvious error. For example, entering 560 instead of 550 is unlikely to cause an out-of-control warning.

Samples should be random. The sampling plan should give every member (specimen) of the population the same chance of selection. A Greek myth and an industrial legend show why this is important.

The ancient Greeks sacrificed their best meat to the gods. The god Prometheus felt sorry for the people who were giving up their best food. He went down to the mortals and told them to prepare two altars. He told them to put the choicest meats on one altar and cover them with bones and tendons. He had them put intestines and other innards on the second altar and cover them with rich fat. Then he invited Zeus, the ruler of the gods, to come down and choose his own sacrifice. No one worried about cholesterol in those days, so Zeus took the altar with the fat on it. He looked only at the material on top of the pile, so what he saw wasn't a random sample (Figure 3.32). Prometheus' ruse worked for the time being, although angering Zeus turned out to be a bad idea.

Here's the same story in an industrial setting. A truck delivered drums to a customer's loading dock. The supplier knew that the customer's quality inspectors looked at the drums in the back of the truck. These were easiest to reach when the driver opened the doors. The supplier loaded the truck as shown in Figure 3.33.

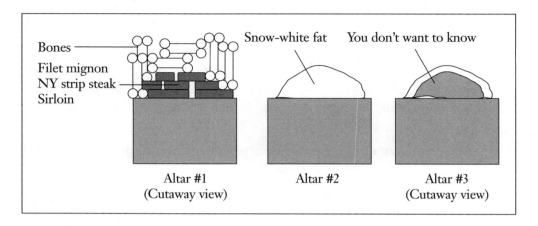

Figure 3.32. Nonrandom sample story from ancient Greece.

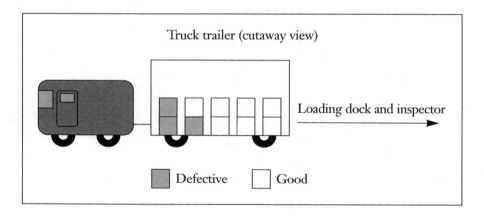

Figure 3.33. Nonrandom sample story, twentieth-century version.

The customer found out about the bad drums later and, like Zeus, was probably angry. This happened many years ago, and customers don't put up with poor quality today. A manufacturing process can, however, do this. Consider a workstation with four positions. The pieces from this workstation travel on a conveyor belt, as shown in Figure 3.34. The ones in front are easiest to reach. What if something is wrong with one of the station's working positions?

Systematic sampling means taking measurements at specific intervals. For example, measuring every fifth unit or group is systematic sampling. It will eventually detect the problem in the conveyor belt example, unless the problem systematically affects every fifth piece. Consider the process' potential to do this before using systematic sampling. Figure 3.35 shows systematic sampling.

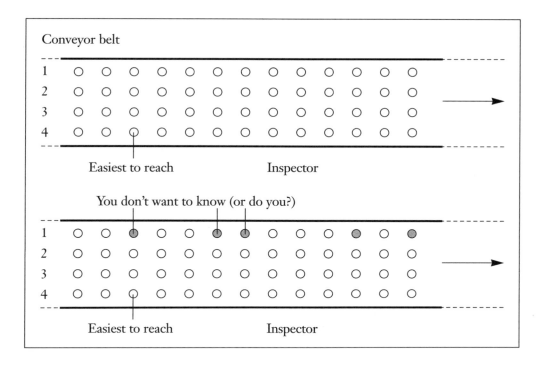

Figure 3.34. Nonrandom sample from manufacturing process.

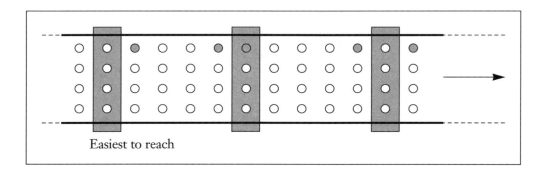

Figure 3.35. Systematic sample from manufacturing process.

Plotting the Data on a Chart

In Figure 3.35 we have three samples of $n = 4$ measurements. If we have a computer, it will do the following calculations for us and plot the results on control charts. If not, we must do the calculations ourselves. Table 3.7 lists the symbols that appear on variables (quantitative) control charts.

Table 3.7. Control chart symbols.

Symbol	Meaning	Explanation
X	Individual measurement	In some manufacturing processes, we can get only one measurement from each sample.
$\bar{x}$ ("x bar")	Sample average	For a sample of n measurements, add the numbers and divide by n.
$\bar{\bar{x}}$ ("x bar bar")	Grand average of all the data	For a total of N measurements, add them and divide by N. If all the sample sizes are equal, $\bar{\bar{x}}$ also is the average of the $\bar{x}$'s. The grand average is an estimate of the actual process mean or center of gravity. It is the centerline of the $\bar{x}$ control chart.
R	Sample range	Subtract the smallest number in the sample from the largest.
$\bar{R}$ ("R bar")	Average range, all samples	This provides an estimate of the process variation.
s	Sample standard deviation	The computer will calculate this.
UCL	Upper control limit	A point above the upper control limit means that the process is out of control.
LCL	Lower control limit	A point below the lower control limit also means that the process is out of control.
CL	Centerline	

A point outside the control limits means that the process is out of control. Assignable or special causes are present. They require investigation and correction.

The sample standard deviation is not practical for hand calculation in a factory environment. The control chart that uses it will probably be on a computer. There are, however, hand calculators that compute standard deviations automatically.

Don't worry about how to compute the control limits. (The procedures for calculating them appear in the appendix.) The chance, however, of exceeding a control limit, if the process is in control, is 0.00135, or 0.135 percent. This is the risk of concluding that something is wrong when the process is working properly. It is the risk of "crying wolf." The normality assumption is important because calculation of these risks relies on it. If the population is not normal, the risks will not be what we think they are.

When there are upper and lower limits, this risk is 2×0.135 percent, or 0.27 percent, or 2.7 per thousand. When the process is working properly, there will be about 2.7 false alarms per 1000 samples. Widening the control limits would reduce the false alarm rate, but would make the chart less sensitive to real problems. Tightening the control limits would make the chart more likely to detect problems, but would cause more false alarms. Increasing the sample size makes the chart more sensitive to real problems without increasing the false alarm rate. Table 3.8 shows a summary.

Now we'll look at an example. Table 3.9 contains simulated data from a silicon oxidation process with a nominal of 1000 Å. The control charts use samples of five measurements. The table includes calculations of the ranges and averages for five samples.

Figure 3.36 shows 50 sample ranges on a range chart. The ranges from the first five samples appear on the chart with the points. For samples of five or less, there is no lower control limit (LCL). (The standard factors for calculating the range chart's LCL are zero for $n = 2$ to 5.) Since this process is in control, no ranges exceed the upper control limit (UCL).

Table 3.8. Sensitivity and false alarms.

Action	Effect on sensitivity to real problems	Effect on false alarm rate (per sample)
Widen the control limits	Reduces sensitivity	Reduces false alarms
Tighten the control limits	Increases sensitivity	Increases false alarms
Take a larger sample	Increases sensitivity	No change

Table 3.9. Process data.

Measurement	Sample 1	Sample 2	Sample 3	Sample 4	Sample 5
1	986.8	1001.9	999.6	1002.0	986.3
2	1010.9	1000.0	993.9	1005.0	993.0
3	994.9	1002.4	998.6	992.0	1009.7
4	991.0	984.5	1008.9	1010.7	998.5
5	1021.9	1005.3	1003.8	1000.9	996.6
Average	5005.5 ÷ 5 = 1001.10	4994.1 ÷ 5 = 998.82	5004.8 ÷ 5 = 1000.96	5010.6 ÷ 5 = 1002.12	4984.1 ÷ 5 = 996.82
Range	1021.9 − 986.8 = 35.1	1005.3 − 984.5 = 20.8	1008.9 − 993.9 = 15.0	1010.7 − 992.0 = 18.7	1009.7 − 986.3 = 23.4

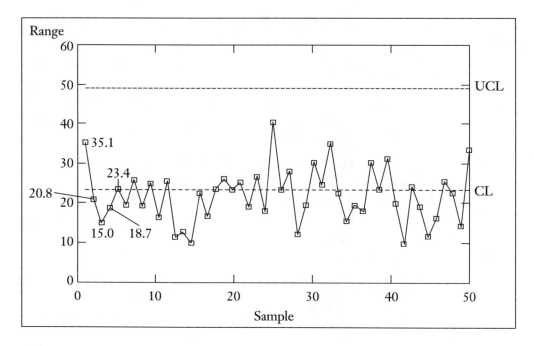

Figure 3.36. Range chart.

Figure 3.37 shows 50 sample averages on a control chart. The averages for the first five samples appear on the chart with the points. Since the process is in control, no points are outside the control limits. The points occur randomly on either side of the centerline.

What happened to the specification limits? There is no relationship between specification limits and control limits. Control limits come from a process characterization study; they depend on the process itself. Specification limits are the customer's requirements.

Specification limits should not even appear on control charts. The terms *in control* and *out of control* refer to the process. Product is in specification or out of specification. A very capable process can make good product (in specification) even when it is out of control. Nonetheless, its chance of making scrap is higher when it is out of control, so it needs correction. An incapable process can make scrap even when it is in control. Trying to adjust or correct this process is futile and may even make it worse. Table 3.10 shows the difference between specification and control.

Note that a sample average can be in specification and in control, while one of the measurements is out of specification. The unit with the out-of-specification measurement is rework or scrap. Do not adjust the process, because it is in control. If, however, the bad piece is an outlier, this suggests an assignable cause.

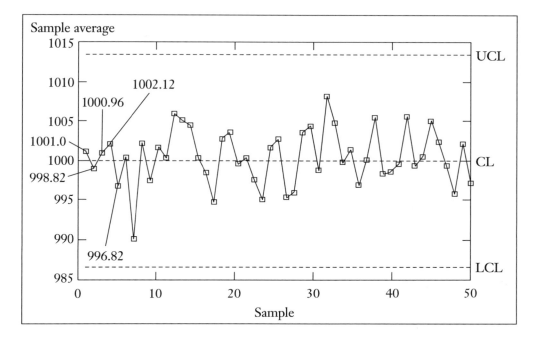

Figure 3.37. $\bar{x}$ chart.

Table 3.10. Specification versus control.

Situation		Product	Process
Out of control, in specification.	USL UCL CL LCL LSL	Ship it; it meets specification.	Requires adjustment; it is out of control.
In control, out of specification.	UCL USL CL LSL LCL	Rework or scrap.	Do nothing. (The process is in control, but not capable; for example a musket.)

For example, let the specification for a 400 Å silicon oxidation process be [360, 440] Å. For a sample of four, the control limits are [385, 415] Å. A sample has the following measurements: 350, 400, 400, 400. The average is 387.5, which is in control. The unit with 350 Å is out of specification and is rework or scrap. The manufacturing team might wonder why one unit was different from the other three.

Also, remember that an outlier strongly affects the range. The range (400 – 350 = 50) may be outside the control limits. Even if it isn't, however, the outlier deserves attention.

Interpreting the Chart

The next step is to interpret the control charts. Look at the range (or standard deviation) chart first. It detects changes in process variation. The control limits for the $\bar{x}$ chart assume that the process variation is constant. If this is not true, we cannot interpret the $\bar{x}$ chart. This does not matter, however, because a change in process variation requires corrective action.

Let's start with a process that is in control. The process nominal is 1000 Å of silicon dioxide. The specifications are [960, 1040], and the standard deviation is 10. Figure 3.38 puts the control chart, target, and histogram side by side. All three have the same meaning.

Figure 3.39 shows what happens if the process variation increases. The R chart or s chart detects changes in process variation.

The chart for sample average is also out of control, but there is no problem with the process mean. The increase in variation makes it more likely

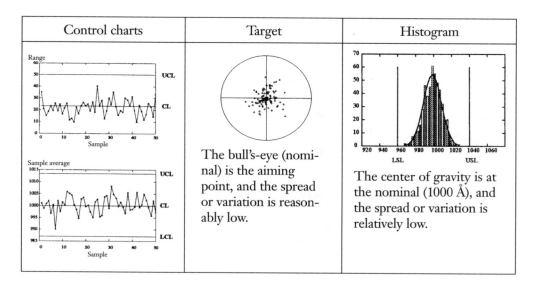

Control charts	Target	Histogram
	The bull's-eye (nominal) is the aiming point, and the spread or variation is reasonably low.	The center of gravity is at the nominal (1000 Å), and the spread or variation is relatively low.

Figure 3.38. Control charts for in-control process.

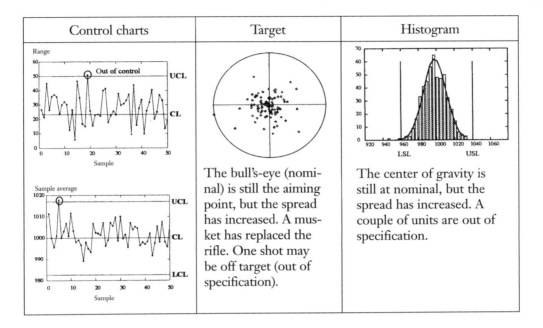

Figure 3.39. Control charts after increase in process variation.

that a sample average will exceed a control limit. In this situation, we cannot meaningfully interpret the $\bar{x}$ chart. When both the sample average and variation charts are out of control, variation is the likely cause.

The $\bar{x}$ chart's function is to detect changes in the process mean. If the variation chart is in control, the next step is to interpret the $\bar{x}$ chart. Figure 3.40 shows a shift in the process mean.

Although the process was out of control throughout the simulation, only one point exceeded the UCL. Being inside the control limits does not prove that the process is in control, just like acquitting a defendant doesn't prove his or her innocence. It means there is not enough evidence to be reasonably sure that the process is out of control. A common mistake in reacting to a point outside the limits is to watch the next run and see if it comes back into control. The out-of-control point is a warning that something is wrong.

This process was out of control before a point exceeded the upper limit. Could we have detected the problem sooner? There are additional tests we can perform on the chart for averages. These are the Western Electric zone tests, which came from the Western Electric Company, which is now part of AT&T (see Table 3.11). (Similar tests are available for the range and standard deviation charts; however, they are more complex.) These tests are optional. Every extra test, while increasing the chance of detecting a problem, also increases the false alarm rate. Harris Semiconductor's Mountaintop plant adds only the zone C test to the basic control limit test.

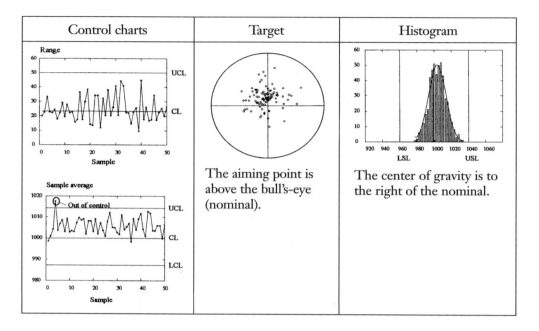

Control charts	Target	Histogram
	The aiming point is above the bull's-eye (nominal).	The center of gravity is to the right of the nominal.

Figure 3.40. Control charts after increase in process mean.

Table 3.11. Western Electric zone tests.

The chart for averages signals an out-of-control condition (shift in process mean) if:		
Test	Situation	Chance of false alarm (upper and lower zones)
Control limits (basic)	One point exceeds either control limit.	2.7 per thousand samples
Zone A	Two out of three successive points in zone A+ or A–. (Read *or* as an exclusive *or*: One in A+ and one in A– is not a failure. The points must be in the same section of the control chart.)	3 per thousand
Zone B	Four out of five successive points in zone B+ or zone B–. (A+ counts as B+, and A– counts as B–.)	5.4 per thousand
Zone C	Eight successive points on one side of the centerline.	7.8 per thousand

Zone C is the easiest to understand. Start by dividing the control chart into six equal zones between the control limits, as shown in Figure 3.41. If the process mean is on the centerline, there is a 50 percent chance that a sample average will fall above the centerline. This is the same chance of throwing a fair coin and getting heads. Eight points on one side of the centerline is like throwing a coin and getting eight successive heads or tails. The chances of doing this are 1 in 128, or 7.8 in a thousand. If you did this, you'd probably conclude that something was wrong with the coin. The same conclusion applies to the process for eight successive samples on one side of the centerline. The principle behind the other tests is similar, but not as easy to explain. Figure 3.42 shows application of the zone tests.

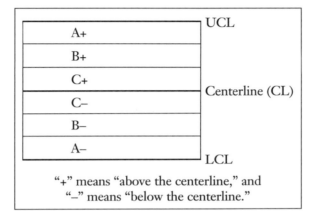

Figure 3.41. Control chart zones.

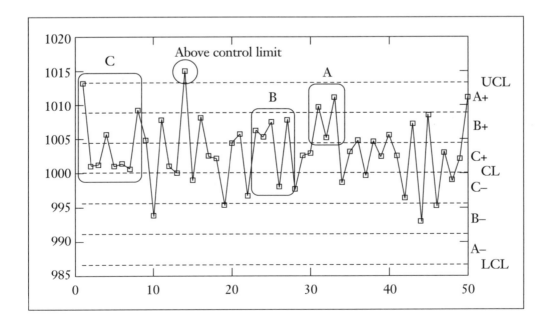

Figure 3.42. Application of the Western Electric zone tests.

Special Patterns

The patterns in Figure 3.43 are characteristic of special situations.

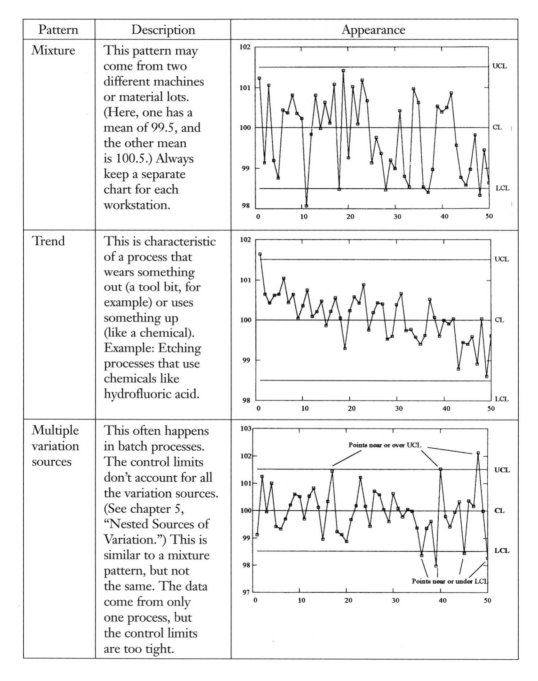

Pattern	Description	Appearance
Mixture	This pattern may come from two different machines or material lots. (Here, one has a mean of 99.5, and the other mean is 100.5.) Always keep a separate chart for each workstation.	
Trend	This is characteristic of a process that wears something out (a tool bit, for example) or uses something up (like a chemical). Example: Etching processes that use chemicals like hydrofluoric acid.	
Multiple variation sources	This often happens in batch processes. The control limits don't account for all the variation sources. (See chapter 5, "Nested Sources of Variation.") This is similar to a mixture pattern, but not the same. The data come from only one process, but the control limits are too tight.	

Figure 3.43. Special patterns.

Pattern	Description	Appearance
Stratifi-cation	This is too good to be true. The control limits are too loose. If a process improvement reduced the process variation, the control limits need recalculation to reflect this.	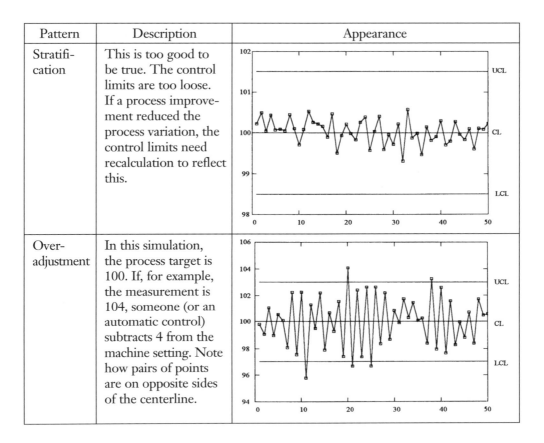
Over-adjustment	In this simulation, the process target is 100. If, for example, the measurement is 104, someone (or an automatic control) subtracts 4 from the machine setting. Note how pairs of points are on opposite sides of the centerline.	

Figure 3.43. *(continued).*

Problems

Problem 1. The SPC charts for this process use a sample of five. Use the basic control limits and the zone C test for the $\bar{x}$ chart (eight points above or below the centerline).

	Lower	**Upper**
Specification limit	7.7 microns	8.3 microns
Control limit, range chart	none	0.492 microns
Control limit, $\bar{x}$ chart	7.87 microns	8.13 microns

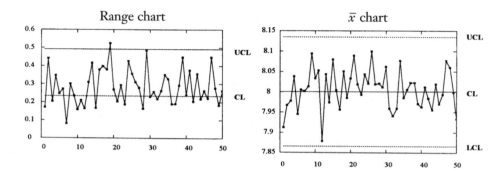

a. Did this process need investigation and corrective action?

b. If so, when should this have happened?

Problem 2. The SPC charts for this process use a sample of five. Use the basic control limits and the zone C test for the $\bar{x}$ chart (eight points above or below the centerline).

	Lower	**Upper**
Specification limit	7.7 microns	8.3 microns
Control limit, range chart	none	0.492 microns
Control limit, $\bar{x}$ chart	7.87 microns	8.13 microns

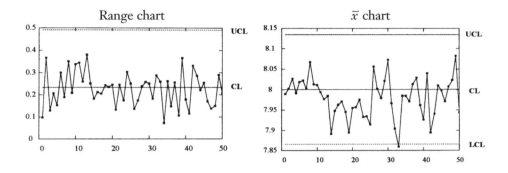

a. Did this process need investigation and corrective action?

b. If so, when should this have happened?

Problem 3. The SPC charts for this process use a sample of nine. The manufacturing team introduced a process improvement to reduce variation. Use the basic control limits and the zone C test for the $\bar{x}$ and R charts (eight points above or below the centerline). For samples of nine, this zone test is reasonably accurate for the range chart.

	Lower	**Upper**
Specification limit	7.7 microns	8.3 microns
Control limit, range chart	0.055 microns	0.539 microns
Control limit, $\bar{x}$ chart	7.9 microns	8.1 microns

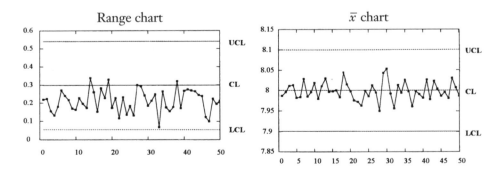

a. Did the process improvement work?

b. If so, when should this have been decided?

c. Why do the points on the $\bar{x}$ chart cluster around the centerline?

Problem 4. The SPC charts for this process use a sample of five. Use the basic control limits and the zone C test for the $\bar{x}$ chart (eight points above or below the centerline).

	Lower	Upper
Specification limit	23.0 mils	27.0 mils
Control limit, range chart	none	2.46 mils
Control limit, $\bar{x}$ chart	24.33 mils	25.67 mils

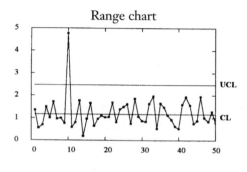

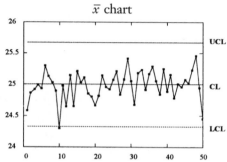

a. Is there a problem with this process?

b. If so, what is a possible explanation?

c. Here is the 10th sample (measurements in mils). What action, if any, is necessary?

$$25.06 \quad 24.83 \quad 24.82 \quad 21.00 \quad 25.78$$

Problem 5. The SPC chart for this process uses a sample of one. This is a chart for individuals (X chart), and there is no range chart. Use the basic control limits and the zone C test for the X chart (eight points above or below the centerline).

	Lower	Upper
Specification limit	98 mm	102 mm
Control limit, X chart	97 mm	103 mm

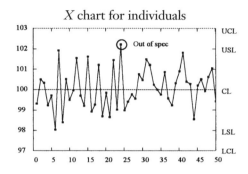

a. What action, if any, should be taken on the process?

b. What action, if any, should be taken on the 24th piece?

Solutions

1. The 18th range exceeded the UCL. The manufacturing team should have investigated the process for an increase in variation.

2. The 18th sample average was the eighth point in a series below the centerline. This means that the process mean has decreased. The 33rd point is below the LCL, but investigation was necessary for the 18th.

3. (a and b) The improvement worked, since the process variation has decreased. For samples of nine, the zone C test (run of eight) works for the range. There is a run of eight consecutive points below the range chart's centerline. This happened on the eighth point. (c) Recall that the $\bar{x}$ chart's control limits depend on the estimate for the process variation. The process' variation was higher before the process improvement. Therefore, the control limits are wider than they should be for the new (lower) variation level.

4. (a) Both the range and $\bar{x}$ chart are out of control. (b) Recall that outliers exert a lot of leverage on the sample average and strongly affect the range. This situation is characteristic of samples with outliers, although outliers are not the only cause. (c) The 21 mil measurement does not look like it belongs with the other measurements. Consider repeating the measurement. If it is still close to 21 mils, there is something wrong with the process. The piece itself is scrap or rework, since it does not meet the specifications.

5. Since the process is in control, adjusting it will only make it worse. The process is not capable, and the only way to get better results is to improve it. The 24th piece is rework or scrap.

CHAPTER FOUR

Attribute Control Charts

Anything we must measure with whole numbers is an attribute. Attributes include nonconformances (rejects, rework, and scrap) and nonconformities (defects). Variables (quantitative or numerical measurements) are preferable, since they provide more information than attributes. Sometimes, however, the process does not yield quantitative data. This chapter shows how to get the most out of attribute data.

We previously discussed control charts for variables or quantitative measurements. Anything we must measure with whole numbers is an attribute. If we can measure it on a continuous scale like a ruler, it is a variable. For example, a part can be 102.1 mils thick, or a feature can be 15.02 microns wide. In contrast, we can have 10 rejects or 11 rejects, but not 10.5 rejects. A part can have no defects or five defects, but it can't have 4.95 defects. Figure 4.1 compares attribute and variable data.

Attributes include *nonconformities* (defects) and *nonconformances* (rework and scrap). A part can have more than one nonconformity, but it can only be one nonconformance. Figure 4.2 shows that a defect may or may not make a part nonconforming, depending on the specifications.

Gages that classify parts as good or bad are "go/no-go" gages. The data are binary: yes/no, 1/0, pass/fail, or good/bad. These are the least useful data we can get. Sometimes, however, they are the only data we can get. This chapter discusses how to use these data as intelligently as possible.

In decreasing order of usefulness, we have the following:

1. Quantitative or variables data (numerical measurements)

   a. Samples, or multiple measurements, are better than individual measurements.

      1. We cannot use a range or standard deviation control chart for individuals.

      2. Large samples make the control chart more powerful or better able to detect process shifts.

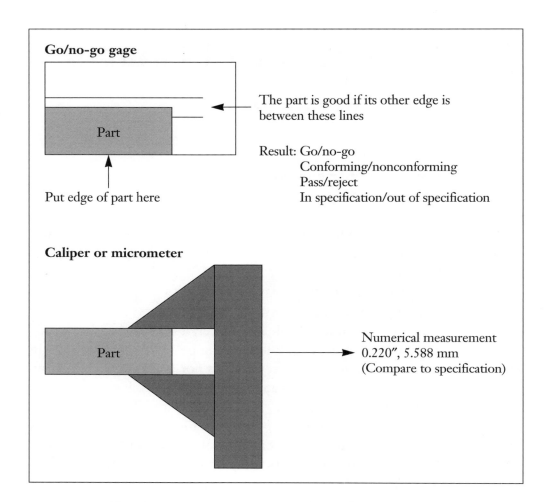

Figure 4.1. Attribute data versus variable (quantitative) data.

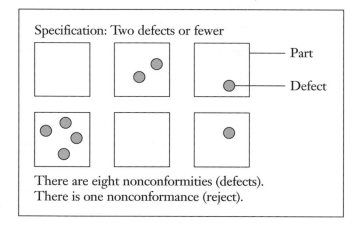

Figure 4.2. Nonconformities and nonconformances.

2. Attributes: nonconformity (defect) data

   a. Defect counts provide more information than simple good/bad counts.

3. Attributes: nonconformance (good/bad, pass/fail) data

   a. Go/no-go data are not effective for controlling, or even identifying, high-capability processes. These are processes whose reject rates are in parts per million.

   b. Attributes data are better than nothing.

4. No data

Here is why quantitative data are much better than attribute data. Consider a manufacturing process that makes 0.00633 percent (63.3 ppm) scrap. Suppose a process shift changes the scrap rate from 0.00633 percent to 2.275 percent. This is a 360-fold increase.

A traditional or Shewhart control chart for individual measurements has a 15.9 percent chance of detecting this. That is, every individual measurement has a 15.9 percent chance of telling us there is a problem. This assumes we use only the control limits and none of the zone tests. To get the same sensitivity from attributes, we'd have to put eight pieces through a go/no-go gage. (We assume the process is good if there are no bad pieces and that there is trouble if one or more are bad.) In this example, one numerical measurement is as useful as eight go/no-go tests.*

---

*Mathematical explanation:

A normally distributed manufacturing process whose specifications are four standard deviations from its mean will make 0.00633 percent rejects. If the process mean shifts two standard deviations, the reject rate will be 2.275 percent. The mean will be one standard deviation from a control limit. There is a 15.9 percent chance that each measurement will exceed the control limit. (Chapter 6, which discusses the cumulative normal distribution, shows how to compute this.)

If the scrap rate is 2.275 percent, each piece has a 0.97725 chance of being good. The chance of eight pieces being good is $0.97725^8 = 0.832$, or 83.2 percent. The chance that the sample will reveal a problem is 16.8 percent, or slightly more than 15.9 percent. (A sample of seven would have a 15 percent chance of detecting the problem, or less than 15.9 percent. $0.97725^{7.5}$ is about 84.1 percent, which yields a 15.9 percent chance of detection, but 7.5 pieces can't be tested.)

Here are the equations that went into creating Figure 4.3.

Nonconforming portion (> upper spec): $p = 1 - \Phi\left(\dfrac{\text{USL} - \mu}{\sigma}\right)$ where USL is 4 in this example, μ is the process mean, and $\sigma = 1$. Φ is the cumulative standard normal distribution.

Chance of nondetection, Shewhart chart: $\Phi\left(\dfrac{\text{UCL} - \mu}{\sigma / \sqrt{n}}\right)$ $\quad \text{UCL} = \dfrac{3}{\sqrt{\text{sample size}}}$

Chance of nondetection, go/no-go sample, pass on 0, fail on 1: $(1 - p)^n$

For equal chances of detection (powers), let: $\Phi\left(\dfrac{\text{UCL} - \mu}{\sigma / \sqrt{n}}\right)$ $(1 - p)^n$, $n = \dfrac{\ln\Phi\left(\dfrac{\text{UCL} - \mu}{\sigma / \sqrt{n}}\right)}{\ln(1 - p)}$

Figure 4.3 shows the go/no-go sample size that has the same power as an X chart for detecting reject rates. For example, suppose the reject rate increases to 0.9 percent. A go/no-go sample of 10 has the same chance of detecting this as a single measurement on an X chart. This assumes that we accept the process if all 10 pieces are good and call it out of control if any are bad.

A process yield* or a defect density is an attribute. The yield is the ratio of good parts to the total. For example, if we make 100 pieces and get 90 good ones, the yield is 0.90 or 90 percent. Although 0.90 is not a whole number, it is the ratio of whole numbers. We treat a ratio of whole numbers as an attribute. The defect density is the portion of defects in a sample. If there are three defects among 200 parts, the defect density is 3/200 or 0.015. Again, 0.015 is the ratio of two whole numbers, and it is an attribute. Table 4.1 shows examples of variables and attributes.

---

*In continuous processes such as chemical manufacture, the yield is a continuous-scale number (variable). This discussion refers to good pieces out of total pieces. The defect density is x defects divided by n pieces (x / n).

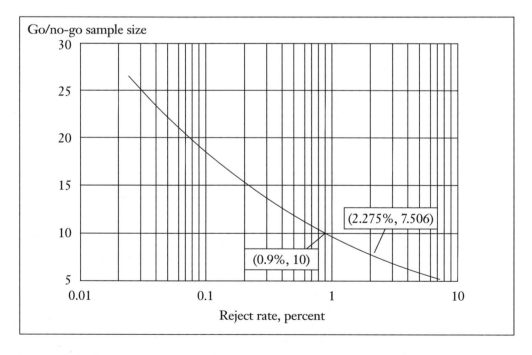

Figure 4.3. Go/no-go sample to detect reject rates (same sensitivity as X chart).

Table 4.1. Variables and attributes.

Variables (quantitative data)	Attributes

• Dimensions
  • Length, width, thickness
• Temperature
• Pressure
• Relative humidity
• Electrical resistivity
  • Also current, voltage
• Viscosity
• Hardness
• Tensile strength
• Chemical concentration
• Impurity levels, for example, ppm
	• Good/bad
	• Rework and scrap
	• Process yield (good ÷ total)
	• Defect counts
	• Defect densities
	• Particle counts
	  • Such as in semiconductor cleanroom

Traditional Attribute Control Charts

Table 4.2 describes the traditional control charts for attributes. The np and p charts are for nonconformances, while the c and u charts are for nonconformities. The np and c charts are for fixed sample sizes, while the p and u charts are useful when the sample size can vary. (The control limits for these charts vary with sample size.)

np *and* p *Charts*

The np and p charts detect changes in rework or scrap rates. They usually detect an increase, but they also can show whether a process improvement worked.

Consider a process that makes, on average, 2 percent rejects. Here are the control limits for a sample of 200.

$$200 \times 0.02 \pm 3\sqrt{200 \times 0.02 \times (1 - 0.02)} = [-1.94, 9.93]$$

These need some changes, since we can't have fractional or negative rejects. If there are fewer than 9.93 rejects, we assume that the process is in control. If there are more, we assume it is out of control. Therefore, the UCL is 9 and the LCL is zero. Figure 4.4 shows the np chart.

The centerline is simply the number of bad pieces we expect. In this case, it's 200×0.02, or 4. A point above the UCL means the reject rate has probably increased.

Table 4.2. Traditional attribute control charts.

Chart	Data plotted	Application	Control limits
np	Nonconformance (reject, scrap, rework) count, np	Shows whether the nonconformance rate has changed.	$n\bar{p} \pm 3\sqrt{n\bar{p}(1-\bar{p})}$, where $\bar{p}$ is the average nonconforming fraction, and n is the sample size.
p	Nonconformance fraction, or percent, p	Same.	$\bar{p} \pm 3\sqrt{\dfrac{\bar{p}(1-\bar{p})}{n}}$
c	Defect count	Shows whether the average defect rate has changed.	$\bar{c} \pm 3\sqrt{\bar{c}}$, where $\bar{c}$ is the average defect count per sample.
u	Defect density (defects per part)	Same.	$\bar{u} \pm 3\sqrt{\dfrac{\bar{u}}{n}}$, where $\bar{u}$ is the average defect density.

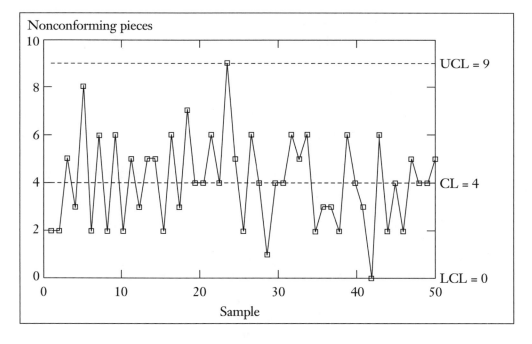

Figure 4.4. np chart.

Figure 4.5 shows the same data in the form of a *p* chart. (Simply divide the ordinate by 200.) The charts look the same, so why use one instead of the other? The *p* chart's scale does not depend on the sample size, while the *np* chart's does. The *p* chart is customary when the sample size varies. When the sample size varies, the control limits also vary. The section on *u* charts for nonconformities includes an example of varying control limits

c and u Charts

The *c* and *u* charts detect changes in defect rates. They also apply to situations that involve random arrivals, like particle concentrations in semiconductor cleanrooms.

How do defects differ from nonconformances? (1) A part can have more than one defect, but it can only be one nonconformance; and (2) A defect (nonconformity) may or may not make the part nonconforming (rework or scrap). This depends on the specifications.

The *c* chart is like the *np* chart. It plots the number of defects in each sample. The *c* chart requires a constant sample size.

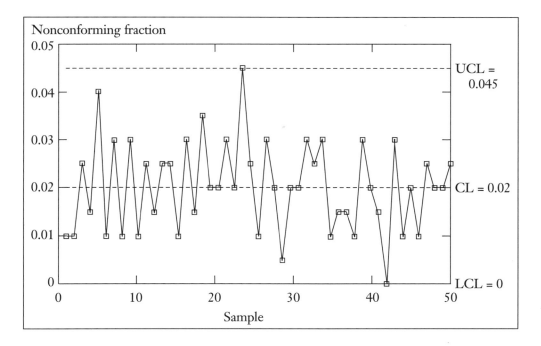

Figure 4.5. *p* chart.

The *u* chart is like the *p* chart. It plots the defect density or defects per sample (*u* = defects ÷ pieces). The *u* chart accepts varying sample sizes, and its control limits depend on the sample size. Table 4.3 shows data for a *u* chart for a process whose average defect density is 0.02 (2 percent). Figure 4.6 shows the chart.

Points Below the Lower Control Limit

Points below the LCL suggest that (1) An experiment to improve the process was successful; or (2) The inspection or test is not detecting nonconformances or defects.

Remember that attribute data measure undesirable events like rework, scrap, and defects. A point outside the control limits suggests the presence of an assignable cause. Assignable causes usually make the process worse,

Table 4.3. Data for *u* chart ($\bar{u} = 0.02$).

Lot	Sample size	Defects	*u*	LCL	UCL
1	450	13	0.0289	0.0000	0.0400
2	450	6	0.0133	0.0000	0.0400
3	450	6	0.0133	0.0000	0.0400
4	450	13	0.0289	0.0000	0.0400
5	300	6	0.0200	0.0000	0.0445
6	300	5	0.0167	0.0000	0.0445
7	300	7	0.0233	0.0000	0.0445
8	600	16	0.0267	0.0027	0.0373
9	600	21	0.0350	0.0027	0.0373
10	600	8	0.0133	0.0027	0.0373
11	450	12	0.0267	0.0000	0.0400
12	450	12	0.0267	0.0000	0.0400
13	450	9	0.0200	0.0000	0.0400
14	450	7	0.0156	0.0000	0.0400
15	300	5	0.0167	0.0000	0.0445
16	300	8	0.0267	0.0000	0.0445
17	300	7	0.0233	0.0000	0.0445
18	600	6	0.0100	0.0027	0.0373
19	600	18	0.0300	0.0027	0.0373
20	600	10	0.0167	0.0027	0.0373

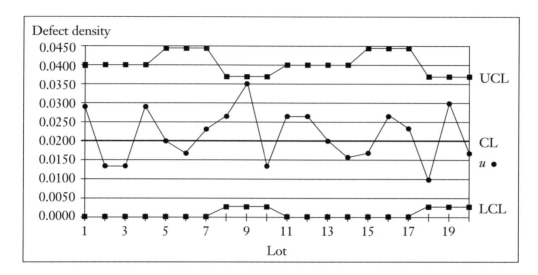

Figure 4.6. *u* chart.

so they're unlikely to reduce rejections or defects. Assignable causes are not likely to make the rejection or defect count go below the LCL. What should we suspect if there is a point below an attribute control chart's LCL?

There is, of course, a false alarm risk for getting a point below the LCL when the process is in control. If we tried an experiment to improve the process, a point below the LCL suggests that it worked. There are really fewer rejections or defects. If this is true, we must recalculate the control limits to reflect the improvement. This recalculation makes the limits tighter and more stringent. It is part of holding the gains (step 7 of TOPS-8D), and making the improvement permanent.

A less pleasant possibility is that the inspection or test has stopped detecting nonconformances or defects. This is very dangerous, since it means we are shipping defective parts. Usually, however, a gage malfunction will cause the opposite problem. The test will reject good parts or find defects where none exist. Figure 4.7 shows a schematic diagram of an electrical test in microelectronics manufacturing. If the electrical probe tips miss the contact pads, there will be an electrical open. This will cause a failure and will certainly not pass a unit that should have failed. In either case, this consideration reemphasizes the need for routine gage calibration.

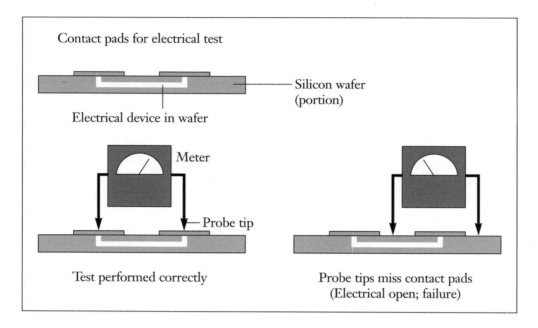

Figure 4.7. A gage malfunction will probably cause rejections.

Normal Approximation (Assumption)

When a manufacturing process is in control, nonconformances follow the *binomial statistical distribution*. This describes samples from large, homogeneous populations whose members are either good or bad. It is standard practice to treat a manufacturing process as an infinite population of parts.

Nonconformities or defects follow the *Poisson distribution*, which describes random arrivals. When the expected number of nonconformances or defects is large, these distributions behave like normal (bell-shaped) distributions. The attribute control charts rely on this assumption.

Let's have *event* mean a nonconformance or a defect. The normal approximation is reasonably good when we expect four or more events (ASTM 1990, 58–59). In our *np* chart example, we expect four nonconformances in each sample. Therefore, the traditional *np* chart is appropriate.

The normal approximation becomes very unreliable when we expect one or fewer events in each sample. Do not use the *np, p, c,* or *u* chart in these situations. Figures 4.8 and 4.9 show the binomial and normal distributions when we expect four and one rejects respectively. Even when we expect four, the normal distribution shows a small chance of getting negative rejects. In a practical situation, this is, of course, impossible. Remember that the normal distribution only approximates the binomial. When we expect only one reject, the approximation is much worse.

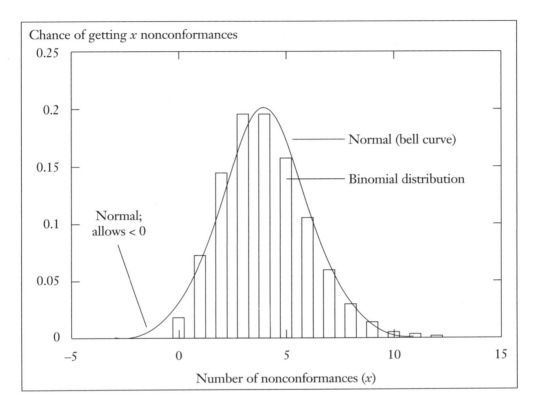

Chance of getting *x* nonconformances

Number of nonconformances (*x*)

Figure 4.8. Normal approximation to binomial distribution, expecting four rejects.

Multiple Attribute Control Charts

A multiple attribute control chart is a check sheet or tally sheet with control limits. Multiple attribute control charts are better than traditional attribute charts when independent problem sources affect a process.

- Multiple attribute charts provide frequency data for Pareto charts.
- Their out-of-control signals point to the problem source.
- They are more likely to detect a problem if one is present.
- They are easier to use than traditional charts.

The traditional control charts (*np*, *p*, *c*, and *u*) count nonconformances and defects. They don't look at why the part is a nonconformance or which defects are present. We discussed check sheets for counting occurrences of different problems. We looked at Pareto charts, which sort problem causes by frequency. Traditional attribute charts lose the advantages of the check sheet and Pareto chart. Harris Semiconductor has found that multiple

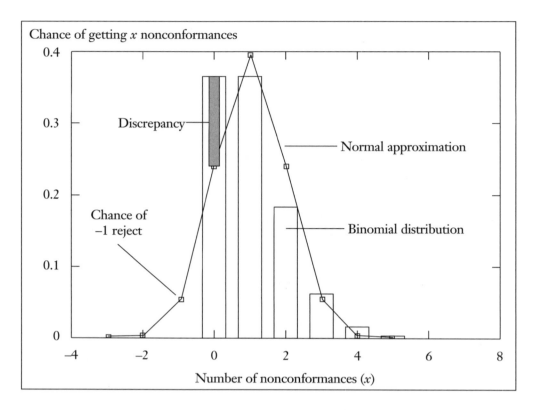

Chance of getting x nonconformances

Discrepancy

Normal approximation

Chance of
−1 reject

Binomial distribution

Number of nonconformances (x)

Figure 4.9. Normal approximation to binomial distribution, expecting one reject.

attribute control charts exceed the capabilities of the traditional charts, and synergize with the tally sheet and Pareto chart. *Multiple attribute* simply means that the chart recognizes and tracks each problem source separately. Cooper and Demos (1991) and Levinson (1994) discuss multiple attribute charts.

Figure 4.10 shows samples of 50 with nonconformances from different sources. Under a traditional charting system, we would report five rejects from the first sample and nine from the second sample. The multiple attribute system would break these down by causes. It might show, for example, that problem B has become worse. The traditional chart could easily miss this.

Figure 4.11 shows a multiple attribute control chart for three reject causes. It accepts a 1 percent risk for false alarms above the UCL for each problem. That is, there is about a 3 percent false alarm risk for each sample. Problems A and C have historically caused 0.5 percent rejects, and B has caused 1 percent. The control limits for A and C are [0, 4]. For B, they are

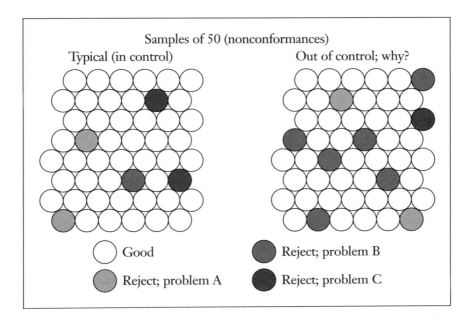

Figure 4.10. Nonconformances from different sources.

		Nonconformance type, historical fraction			
		A	B	C	Total
		0.50%	1.00%	0.50%	2.00%
	UCL	4	7	4	10
	LCL	0	0	0	1
Lot	Sample				
1	250	0	2	1	3
2	250	1	0	1	2
3	250	2	3	0	5
4	250	0	1	3	4
5	250	0	7	2	9
6	250	2	3	2	7
7	250	1	2	6 H	9
8	250	0	0	1	1
9	250	2	2	4	8
10	250	2	5	2	9
Total	2500	10	25	22	57
Average		0.40%	1.00%	0.88%	2.28%

Figure 4.11. Multiple attribute control chart (constant sample size).

[0, 7], and for the total, they are [1, 10]. The false alarm risk for the total (A + B + C) is 3 percent. For comparison purposes, consider the following:

Option 1: Track A, B, and C separately, with a net false alarm risk of 3 percent.

Option 2: Track (A + B + C), with a 3 percent false alarm risk.

The important features of the multiple attribute chart include the following:

- It works exactly like a check sheet or tally sheet. For each sample, the inspector enters the number of each reject type.

- The "total" row provides the information for a Pareto chart. The totals for A, B, and C are 10, 25, and 22 respectively.

  —The chart also lets us compute the average nonconformance rate for each problem. The example assumes that historical information is available. If we did not have this information, we would need to get it from a check sheet.

- If an entry exceeds a control limit, highlight it.

  —The figure comes from a Microsoft Excel spreadsheet that does this automatically. If the entry is above the UCL, a red *H* appears for high. If it is below the LCL, a green *L* appears for low.

  —Corel Quattro Pro and Lotus 1-2-3 probably have the same capabilities.

  —Cooper and Demos (1991) highlight entries above the UCL with red boxes and below the LCL with green circles. The two symbol shapes prevent confusion if color is unavailable or if a person is color-blind.

- In this example, the actual rate for C rose to 1 percent for the sixth through tenth samples. The entry for C is above the UCL on the seventh sample. The reject total did not exceed its UCL of 10. The conventional control chart that merely counts total nonconformances would not have detected the problem, at least not by the tenth sample (See Figure 4.12).

  —The UCL of 10 gives the "total" control chart a 3 percent false alarm rate. This is the same chance of getting a false alarm from one of the three problem causes.*

---

*The false alarm risk for each individual problem is nominally 1 percent. If the process is in control, there is a 99 percent chance of getting no false alarm for each problem. The chance of no alarm for all three is 0.99^3 or 0.970 (97.0 percent), so the chance of at least one false alarm is 3 percent. Since we are dealing with attribute sampling plans, however, the risk will rarely equal the nominal risk. The technical appendix shows how to compute the control limits to meet specified false alarm risks.

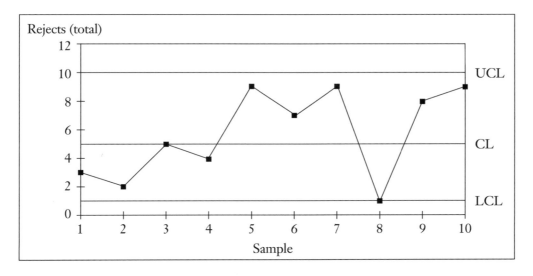

Figure 4.12. Conventional control chart for reject totals (3 percent false alarm risk).

Figure 4.13 shows what happens when the samples are different sizes. The control limits depend on the sample size. This means we must compute control limits for each new sample. This is not convenient to do by hand, but a spreadsheet can easily handle the job. Let the nonconformance rate for problem C again increase from 0.5 percent to 1 percent for the sixth through tenth samples. In Figure 4.13, the spreadsheet calculates the chance of getting six or more rejects for C if the reject rate is really 0.5 percent. There is less than a 1 percent chance of this, so the program flags the entry as out of control (high). (The chart for totals also detects the problem, only because B happened to have six rejects at the same time as C.) We are still using a 1 percent false alarm rate for each problem and a 3 percent rate for the totals. These are the chances of exceeding the UCLs if the process is in control. Similar charts can be used for defect (nonconformance) data.

Technical Appendix: Multiple Attribute Charts

Readers can skip this section without losing continuity. It provides the technical details for using multiple attribute charts.

Hypergeometric Distribution

The hypergeometric distribution applies to samples from small populations without replacement. For example, the hypergeometric distribution defines the chance of getting two aces if we draw five cards from a deck of 52.

Lot	Sample	Nonconformance type, historical fraction							
		A		B		C		Total	
		0.50%		1.00%		0.50%		2.00%	
1	250	2		2		2		6	
2	250	0		4		0		4	
3	250	2		1		0		3	
4	250	0		2		1		3	
5	200	1		2		3		6	
6	200	0		0		3		3	
7	300	0		2		2		4	
8	300	1		6		6	H	13	H
9	300	0		3		3		6	
10	300	1		4		4		9	
Total	2600	7		26		24		57	
Average		0.27%		1.00%		0.92%		2.19%	

Figure 4.13. Multiple attribute chart for variable sample sizes.

Without replacement means that we do not return cards to the deck after we draw them. The same principle applies to drawing a sample of n pieces from a lot of N pieces.

The hypergeometric distribution does not apply to SPC, which assumes that a manufacturing process represents an infinite population. The binomial distribution covers attribute samples from infinite populations. Also, the binomial distribution is an accurate approximation when we sample from a finite population (like a lot), if $n << N$.

Suppose that, in a lot of N pieces, M are defective. The hypergeometric distribution describes the chance ($\Pr(x)$) of getting x defective pieces in a sample of n.

$$\Pr(x) = \frac{\dfrac{M!}{x!(M - x)!} \dfrac{(N - M)!}{(n - x)!((N - M) - (n - x))!}}{\dfrac{N!}{n!(N - n)!}} \quad \textbf{(Eq. 4.1)}$$

Again, we will not use this equation for SPC. It applies to some quality sampling applications, where n is not much smaller than N. It is here primarily for reference.

Binomial Distribution

The binomial distribution applies to samples (1) from small populations with replacement, and (2) from large populations without replacement. Consider the chance of drawing two aces in a sample of five cards. If we return each card we examine to the deck and reshuffle before drawing the next, the binomial distribution applies. This is an example of sampling with replacement. Again, a manufacturing process represents an infinite population, even if the run is finite.

This is why casinos use several card decks for blackjack. If they used single decks, players who can remember which cards were played would gain an advantage. Selection of a card from a single deck reduces the chance of drawing the same card later. A large deck causes the situation to approach sampling from an infinite population without replacement. This prevents anyone from gaining an advantage by counting cards. The alternative would be to reshuffle after each round, but this takes time. (It's like running frequent setups in a factory. The dealer earns money by playing, not shuffling.)

Consider again a finite population of N items, of which M are defective. We take a sample of n and want to know the chance of getting x defective units. Now suppose that $n << N$. Then $(n - x)$ must be much less than $(N - M)$, and $x << M$. Under these circumstances, the hypergeometric distribution approaches the binomial as follows:

$$\Pr(x) = \frac{\dfrac{M!}{x!(M - x)!}\dfrac{(N - M)!}{(n - x)!((N - M) - (n - x))!}}{\dfrac{N!}{n!(N - n)!}}$$

and use
$$\lim_{\substack{n \to \infty \\ n >> x}} \frac{n!}{x!(n - x)!} = \frac{n^x}{x!}$$

$$\lim_{\substack{M \to \infty,\, M >> x \\ (N - M) \to \infty \text{ and} \\ (N - M) >> (n - x) \\ \text{and } N \to \infty,\, N >> n}} \Pr(x) = \frac{\dfrac{M^x}{x!}\dfrac{(N - M)^{n-x}}{(n - x)!}}{\dfrac{N^n}{n!}} = \frac{n!}{x!(n - x)!}\frac{M^x (N - M)^{n-x}}{N^x N^{n-x}}$$

Noting that $p = \dfrac{M}{N}$ and $(1 - p) = \dfrac{N - M}{N}$ we finish with

$$\Pr(x) = \frac{n!}{x!(n - x)!} p^x (1 - p)^{n-x}$$

When a manufacturing process is under control, nonconformances follow the binomial distribution. *Under control* means the nonconformance rate is a constant $100p$ percent. The binomial distribution shows the chance of getting x events out of n trials when this is true.

$$\text{Binomial } \Pr(x|n,\, p) \;=\; \frac{n!}{x!(n-x)!}\, p^x (1-p)^{n-x} \text{ and}$$

(Eq. set 4.2)

$$\Pr(x \le c|n,\, p) \;=\; \sum_{x=0}^{c} \frac{n!}{x!(n-x)!}\, p^x (1-p)^{n-x}$$

$\Pr(x|n,\, p)$ is the binomial probability of getting x events, given n trials and probability p for each trial. The vertical bar means *given* the conditions to its right. The summation formula is the chance of getting c or fewer events out of n trials, if the nonconformance rate is p. The latter is the cumulative binomial distribution.

Let us accept a 1 percent false alarm risk of exceeding the UCL and the same risk of exceeding the LCL. Equation set 4.3 shows how to define the control limits. Note that for an attribute sampling plan, the false alarm risk will rarely equal the specified risk. Equation set 4.3 defines the most sensitive plan whose risk is less than or equal to the specified risk.

$$\text{Find LCL such that } \sum_{x=0}^{LCL-1} \frac{n!}{x!(n-x)!}\, p^x (1-p)^{n-x} \le 0.01$$

(Eq. set 4.3)

$$\text{Find UCL such that } \sum_{UCL+1}^{n} \frac{n!}{x!(n-x)!}\, p^x (1-p)^{n-x} \le 0.01, \text{ or}$$

$$\text{Find UCL such that } \sum_{0}^{UCL} \frac{n!}{x!(n-x)!}\, p^x (1-p)^{n-x} \ge 0.99$$

That is, define the LCL so there is less than a 1 percent chance of going below it if the process is in control. The LCL is the smallest number that makes the cumulative binomial summation greater than 0.01 (1 percent). Similarly, there must be less than a 1 percent chance of exceeding the UCL. The UCL is the smallest number that makes the cumulative binomial summation greater than 0.99 (99 percent).

The example with the fixed sample used a sample of 250. Table 4.4 shows how to find the control limits for binomial processes with $p = 0.005$ and 0.01, and a 1 percent chance of exceeding either control limit. We can

Table 4.4. Binomial control limits.

p	LCL	UCL
0.005	Cum Pr($x \leq 0 \mid 250, 0.005$) = 0.286 (no LCL)	Cum Pr($x \leq 4 \mid 250, 0.005$) = 0.991 UCL = 4. (If the process is in control, there will be four or fewer nonconformances 99.1 percent of the time.)
	= CRITBINOM (250, 0.005, 0.01) where the arguments are the sample size, chance of occurrence, and percentile of the binomial distribution.	= CRITBINOM (250, 0.005, 0.99)
0.01	Cum Pr($x \leq 0 \mid 250, 0.01$) = 0.081 (no LCL)	Cum Pr($x \leq 7 \mid 250, 0.01$) = 0.996, and Cum Pr($x \leq 6 \mid 250, 0.01$) = 0.986. UCL = 7. (If the process is in control, there will be seven or fewer nonconformances 99.6 percent of the time.)
	= CRITBINOM (250, 0.01, 0.01)	= CRITBINOM (250, 0.01, 0.99)
0.02	Cum Pr($x \leq 0 \mid 250, 0.02$) = 0.0064 Cum Pr($x \leq 1 \mid 250, 0.02$) = 0.039 LCL = 1. (There is less than a 3 percent chance of getting zero rejects if n = 250 and p = 0.02.)	Cum Pr($x \leq 10 \mid 250, 0.01$) = 0.987, and Cum Pr($x \leq 9 \mid 250, 0.01$) = 0.969. UCL = 10. (If the process is in control, there will be 10 or fewer nonconformances 98.7 percent of the time. UCL = 9 is too low, because there is a 3.1 percent chance of getting more than nine.)
	= CRITBINOM (250, 0.02, 0.03)	= CRITBINOM (250, 0.02, 0.97)

do the same for p = 0.02 and a 3 percent chance of exceeding either limit. This is a tedious hand calculation, but Microsoft Excel's CRITBINOM function does it automatically.

Note that we are actually using the binomial distribution and not the normal approximation. This means we don't have to worry about whether the normal approximation is good.

What happens when the sample sizes are unequal? We can program the spreadsheet to determine whether each entry is out of control. Figure 4.14 shows how it works for Microsoft Excel. Novell Quattro Pro and Lotus 1-2-3 probably have similar capabilities.

In Figure 4.14, $C$2 contains the two-sided false alarm risk (2 percent). This is the risk of exceeding the LCL or UCL when the process is in control.

	A	B	C	D	E	F	G	H
6			A		B		C	
7	Lot	Sample	0.50%		1.00%		0.50%	
8	1	**250**	2		2		2	= −1*(BINOMDIST **(G8,\$B8,G\$7,**1) < **\$C\$2**/2) + 1* (BINOMDIST (G8-1*(G8 > 0), \$B8,G\$7,1) > 1 − \$C\$2/2)
9	2	250	0		4		0	= −1*(BINOMDIST (G9,\$B9,G\$7,1) < \$C\$2/2) + 1* (BINOMDIST (G9-1*(G9 > 0), \$B9,G\$7,1) > 1 − \$C\$2/2)
10	3	250	2		1		0	
11	4	250	0		2		1	
12	5	200	1		2		3	
13	6	200	0		0		3	
14	7	300	0		2		2	
15	8	300	1		6		6	H
16	9	300	0		3		3	
17	10	300	1		4		4	= −1*(BINOMDIST (G17,\$B17,G\$7,1) < \$C\$2/2) + 1* (BINOMDIST (G17-1*(G17 > 0), \$B17,G\$7,1) > 1 − \$C\$2/2)
18	Total	2600	7		26		24	
19	Average		0.27%		1.00%		0.92%	

Figure 4.14. Portion of Excel spreadsheet for variable sample sizes.

If there is no LCL, the false alarm risk becomes 1 percent. (This is the chance of exceeding the UCL alone.) The arguments of BINOMDIST are BINOMDIST (c, n, p, 1). The summation is from 0 to c; there are n trials (n is the sample size), with probability p. The 1 means to take the cumulative distribution. If this is 0, or omitted, the function returns the chance of getting exactly c events.

Here is what the function is doing in cell H8. The cells below it are similar. There are similar functions in columns D and F for nonconformance types A and B.

–1*(BINOMDIST (G8,$B8,G$7,1) < $C$2/2)	If the cumulative binomial summation from 0 to 2 (cell G8) is less than 1% (2% ÷ 2), put –1 in the cell.
+1*(BINOMDIST (G8 – 1*(G8 > 0), $B8,G$7,1) > 1 – $C$2/2)	If the cumulative binomial summation from 0 to 1 (cell G8 – 1) is more than 99% (100% – 2% ÷ 2), put 1 in the cell. That is, "Is G8 – 1 greater than or equal to the LCL?" (If it is, G8 exceeds it.) G8 – 1* (G8 > 0) prevents the function from choking if G8 is zero. The function is undefined for negative counts.
	If neither of these is true, the cell contains 0.

Finally, the cell format is [< 0][GREEN] "L"; [> 0] [RED] "H";""
If –1 (below the LCL), the cell displays a green L for low.
If +1 (above the UCL), the cell displays a red H for high. (Cell H15 did this.)
If 0 (within the control limits), the cell displays nothing. (No attention is necessary.)

Poisson (Defect or Random Arrival) Situation

When a manufacturing process is in control, defects will follow the Poisson distribution, which applies to random arrivals. The Poisson distribution also describes large samples from binomial systems when the event probability (p) is small. The following derivation relates the two distributions.

1. Binomial $\Pr(x \mid n, p) = \dfrac{n!}{x!(n - x)!} p^x (1 - p)^{n-x}$

2. Find $\lim\limits_{n \to \infty, p \to 0} \dfrac{n!}{x!(n - x)!} p^x (1 - p)^{n-x}$

3. $\lim\limits_{n \to \infty} \dfrac{n!}{(n - x)!} = n^x$ Also, $(1 - p)^{n-x} = \left(1 - \dfrac{np}{n}\right)^{n-x}$

   and $\lim\limits_{n \to \infty} \left(1 + \dfrac{y}{n}\right)^n = e^y$, so $\lim\limits_{n \to \infty} \left(1 - \dfrac{np}{n}\right)^n = e^{-np}$

   Note that when $p \to 0$, $(1 - p)^{n-x} \approx (1 - p)^n$

4. Let np be the Poisson mean μ.

   $\lim\limits_{n \to \infty, p \to 0} \dfrac{n!}{x!(n - x)!} p^x (1 - p)^{n-x} = \dfrac{n^x p^x}{x!} e^{-np} = \dfrac{\mu^x}{x!} e^{-\mu}$

Equation 4.4 shows the Poisson probability of getting x events when the mean (expected number of events) is μ.

$$\Pr(x \mid \mu) = \frac{\mu^x}{x!} e^{-\mu} \qquad \textbf{(Eq. 4.4)}$$

In Microsoft Excel, POISSON $(c, \mu, 1)$ returns the cumulative Poisson summation, $\sum_{x=0}^{c} \frac{\mu^x}{x!} e^{-\mu} = \left(\sum_{x=0}^{c} \frac{\mu^x}{x!} \right) e^{-\mu}$. The 1 tells the function to return the summation instead of the chance of getting exactly c events. There is no CRITPOISSON function that corresponds to CRITBINOM, but the POISSON function lets us set up a chart that will accept varying sample sizes. The Poisson mean for a sample of n and defect density $p = u$ is np. We expect p or u defects per piece. Alternately, we might expect p arrivals per unit time, and np arrivals in n units of time.

Normal Approximation: Binomial and Poisson Distributions

When the binomial or Poisson mean is large, these distributions behave like normal distributions. This allows us to use ±3-sigma (Shewhart) control limits for them. The normal assumption is reasonably good when we expect four or more events (ASTM 1990, 58–59). If a computer is available, we need not use approximations,* but these are useful for hand calculations.

	Binomial	Poisson
Mean	np	$np = \mu$
Variance (σ^2)	$np(1-p)$	μ
Standard deviation	$\sqrt{np(1-p)}$	$\sqrt{\mu}$
Shewhart control limits	$np \pm 3\sqrt{np(1-p)}$	$\mu \pm 3\sqrt{\mu}$
Estimators	For p: $$\bar{p} = \frac{1}{\sum_{i=1}^{m} n_i} \sum_{i=1}^{m} x_i$$ $$= \frac{\text{Total nonconformances}}{\text{Total pieces}}$$ for m samples of size n_i, with x_i nonconformances in the ith sample	$\bar{c}$ is an estimate for μ (c chart) $\bar{u}$, the average defect density, is similar to $\bar{p}$. Instead of x_i nonconformances, use x_i defects in the ith sample.

---

*Note: For large samples, a computer may be unable to calculate $n!$ or $(n - x)!$ in the binomial equation. If you need to program the cumulative binomial, consider a recursive equation, where j ranges from 0 to c, and $\text{BinPr}(0 \mid n, p) = (1 - p)^n$, $\text{BinPr}(j + 1 \mid n, p) = \text{BinPr}(j \mid n, p) \times \left(\frac{n-j}{j+1} \frac{p}{1-p} \right)$.

CHAPTER FIVE

Other Topics

Topics in this chapter include gage capability, inspection capability, and nested sources of variation. It is important to remember that inspections cannot assure extremely high quality.

Gage Capability

If it can't be measured, it can't be controlled. Gages must be accurate (calibrated) and precise (capable) if they are to provide useful information.

We previously discussed calibration, which assures a gage's accuracy. It is, however, only one requirement for effective measurements. The other is precision. We introduced these terms in chapter 3, which discusses a manufacturing process' ability to meet specifications. The mean of an accurate process is on the target, which is usually halfway between the specifications. Precision is the opposite of variation, and it has the same meaning in discussions of gages.

Unlike a manufacturing target, a gage's target is the actual dimension of the specimen it is measuring. If a dimension is 12.0 microns, we want the gage to report 12.0 microns. Table 5.1 shows how a gage is like a manufacturing process. The gage's product (the measurement) has unavoidable variation.

Table 5.1. Similarities between gages and processes.

	Process	Gage
Product	The workpiece	A measurement
Target	Usually the midpoint of the specification	The specimen's actual dimension
Variation	Process variation	Gage variation (reproducibility and repeatability)

An accurate gage will, on average, report the specimen's actual dimension. The words *on average* are unsettling, and they should be. A noncapable (imprecise) gage will return widely differing measurements from the same specimen. If the piece's dimension is slightly out of specification, the gage may pass it. If the piece is slightly within specification, the gage may reject it. Figure 5.1 shows an almost-perfect gage. If a part is bad, it will almost certainly fail. If it is good, the gage will almost certainly pass it. The figure also shows a noncapable gage that can easily pass bad pieces and reject good ones. Figure 5.2 shows the difference between accuracy and precision.

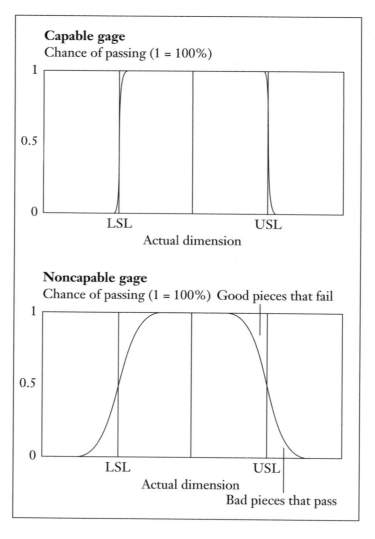

Figure 5.1. Effect of gage precision on quality.

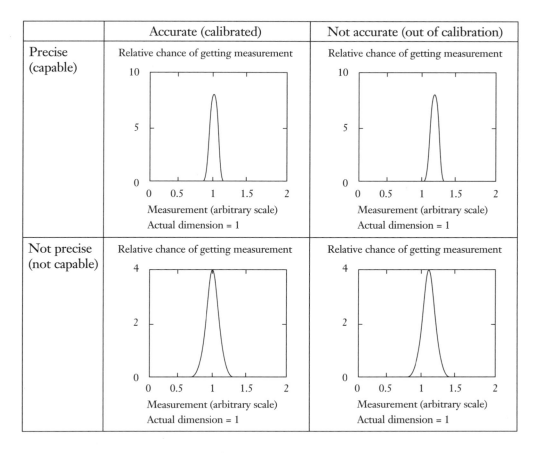

Figure 5.2. Gage accuracy and precision.

Elements of Gage Variation

Reproducibility and repeatability (R&R) are the elements of gage variation. They actually refer to lack of reproducibility and lack of repeatability. If a gage has good *reproducibility*, the measurement will not depend on the person who uses it. If a gage has good *repeatability*, it returns the same number each time a given specimen is measured. A gage study, or R&R study, measures the gage variation.

Reproducibility usually refers to the ability of different operators to reproduce a measurement. We also can look at day-to-day, or shift-to-shift, reproducibility. Does the gage change with time? We could set up a control chart for a gage by measuring a standard specimen every day or every shift. A point outside the control limits for process mean shows that the gage needs recalibration.

Why would different operators get different measurements from the same gage? If the gage requires a lot of subjective judgment or interpretation, its reproducibility will be poor. For example, Figure 5.3 compares a dial meter to a digital display. Dial indicators are notorious for parallax error. Parallax means that the number the user sees depends on his or her viewing angle. Good meters have mirrors behind them to reduce parallax error. By making sure that the needle covers its own reflection, the user assures a 90° (perpendicular) viewing angle. A digital display, however, leaves no room for interpretive error.

Microelectronics manufacturers use microscopes with electronic micrometers, or calipers, to measure features on silicon wafers. The result depends on the user's view of the edge of the feature. Figure 5.4 describes an electronic micrometer.

There is also variation when we measure one specimen several times. Laboratories refer to this variation as experimental error. A gage study, or R&R study, quantifies both variation sources. The total gage variation depends on them both as shown in Figure 5.5.

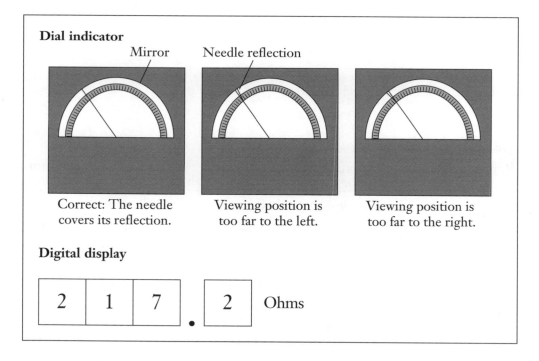

Figure 5.3. Which gage is better?

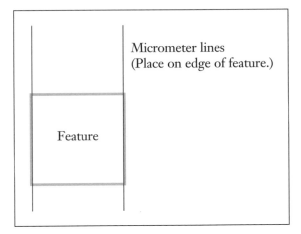

Figure 5.4. Electronic micrometer.

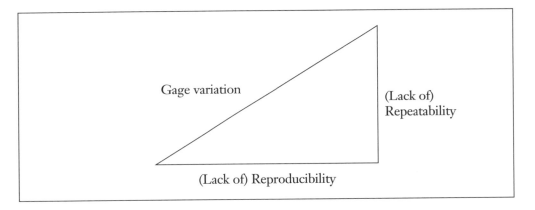

Figure 5.5. Gage variation as a function of repeatability and reproducibility. Each side of the triangle is the corresponding standard deviation (s). The chapter appendix will discuss this in detail.

Hradesky (1988), Barrentine (1991), and Montgomery (1991) describe how to perform gage studies. Although their procedures differ slightly, a gage study returns the percent tolerance consumed by (gage) capability (PTCC), or the percent of tolerance (P/T) ratio. Hradesky (1988, 84) provides guidelines for gage capability (Table 5.2).

Table 5.2. Gage capability guidelines.

PTCC	Status
≤ 10%	Acceptable*
10–25%	Marginal
> 25%	Unacceptable

*The current QS-9000 requirements (section 4.11.4S "Measurement System Analysis") call for PTCC ≤ 10%. The state of the art, however, may limit the availability of gages that meet this goal.

Inspection Capability

We cannot inspect high quality levels into a product. We must build or design quality into a product.

Human inspections are generally 80 percent effective (Juran and Gryna 1988, 18.84). This means that even 100 percent inspections will let about 20 percent of the defects through. Juran and Gryna discuss several reasons for inspection errors.

In the microelectronics industry, the needle in a haystack is a common problem. Inspectors are looking for something that doesn't belong in a complex wiring pattern. It is very easy to miss such defects, especially if they are small.

A theoretically effective, if expensive, way to address this problem is to inspect each piece twice. If another person does the second inspection, his or her different way of looking at the part may catch something the first person missed. If each inspector catches 80 percent of the defects, a defect has a 20 percent chance of getting by each inspector; 20 percent of 20 percent is 4 percent, so only 4 percent of the defects will escape. (A third inspection will reduce this to 0.8 percent.)

There are, however, some practical limitations on this technique. It is costly because we must use twice as many inspectors, or more. Also, the theoretical analysis of multiple inspections assumes that a random 20 percent of the defects escape. In practice, defects that slip through are likely to be smaller or less visible than those that don't. A defect that gets by the first inspector is also likely to escape the second.

Human and morale factors also affect the inspection. People are more alert when they expect something to go wrong. Suppose the leader of a commando squad wants to attack an enemy supply center. Which is the preferred target: the one near the front lines or the one well behind the enemy's lines? (We are assuming equal levels of danger and difficulty in getting to them and escaping afterward.) Of course, this would not be true in practice. The sentries at the front lines will be very alert, since they expect someone to attack them. Those in the enemy's rear areas are not expecting major trouble. At worst, they may worry about thieves stealing the supplies. They may stand their watches because they have to, but they will be far less alert than the frontline sentries. It will be far easier to surprise the rear area sentries.

The first inspector in a multiple inspection is like a frontline sentry. He or she expects something bad and is alert for it. The second inspector is like a rear area sentry. He or she assumes that the first inspector has caught most of the defects and does not expect to find many. This inspector will be less alert. In summary, even multiple inspections cannot inspect quality into the product.

Juran and Gryna (1988) and Hradesky (1988) provide some measurements of an inspection's effectiveness. These measurements require knowledge of the true number of defects that are present. Juran and Gryna (1988, 18.95) define an inspection's accuracy as the percent of defects it identifies correctly.

$$\text{Accuracy} = \frac{d - k}{d - k + b}$$

d = defects reported
k = defects reported incorrectly
(so $d - k$ = defects identified correctly)
b = defects missed by the inspector, so
$d - k + b$ = true defects that were actually present

Hradesky introduces the probability of a miss, or the chance of not detecting a defect. This is simply the reverse of the accuracy, or $b / (d - k + b)$.

When the piece is either good or bad, waste is the percent of good pieces that the inspection rejects. Hradesky calls this the *probability of a false alarm*. Juran and Gryna (1988, 18.97) provide the following expression.

$$\text{Waste} = \frac{k}{n - d - b + k} = \frac{\text{Incorrect rejections}}{\text{Good pieces present}}, \text{ where } n = \text{sample size}$$

Improvements in Inspection

Many decades ago, astronomers developed a way to find the needle in the haystack. They thought there was a planet (Pluto) in a certain area of the sky. To find it, they aimed a large telescope at that area and took photographs. Astronomical photography involves exposing photographic plates for several minutes or longer. A clockwork motor adjusts the telescope to compensate for the earth's rotation while this is happening. The result is a photograph of thousands of stars, most of which are not visible to the human eye. How did the astronomers find Pluto against this background?

From a telescope's viewpoint, stars don't move and planets do. Since Pluto orbits the sun, it moves visibly over several days or weeks. The astronomers took a second picture some time after the first. If a speck of light changed positions, it was a planet. Finding this point among thousands of others was still like looking for a needle in a haystack, but it took only a moment's inspection to find it.

The astronomers placed the two photographs in a blink comparator. This device aligns the photographs and rapidly blinks between them. It shows the first one for an instant, then the second, and so on. The astronomers saw a field of thousands of stars and one point flickering back and forth. The inspection tool took advantage of the eye's inherent attraction to motion.

This procedure could be useful for finding defects in complex patterns. A blink comparator could compare a good (standard) piece against each part that comes to it. Anything that doesn't belong there will flicker on and off, thus attracting the inspector's attention.

Figure 5.6 shows two parts. One has some defects, and the other doesn't. Try the following. Make several photocopies of the figure, and cut out the pictures of the parts. Mount the pictures on cards or other relatively firm paper and alternate the left and right pictures. Then fan the cards to produce a moving picture effect. (Even better, scan the images into a computer graphics program with animation capability. Then superimpose and alternate them.) The defects will "move" during the animation.

Better methods are available today. Machine vision systems compare each part to a standard or golden part. Alternately, inspectors can teach these machines what a good piece looks like. The machine vision system detects pieces that don't match the image of a good part. Machines, unlike people, don't suffer from boredom or inattention. They are consistently alert, even when the defect rate is one in a million. They are not perfect, however, and they lack human judgment.

Designing and Building in Quality

The best way to assure quality is to design or build it into the product. DFM means designing the product to be easy to make. The product designers need to work closely with manufacturing to understand the capabilities of the process. DFM includes design review, which covers factors such as making sure the manufacturing equipment can meet the tolerances. This avoids the common mistake of designing something that a development laboratory can make but a factory can't.

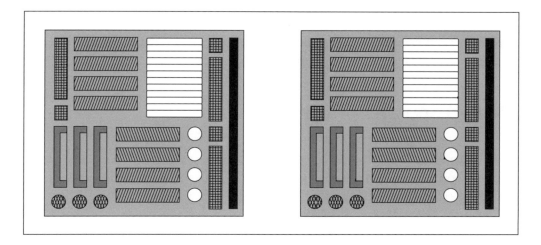

Figure 5.6. Defects on a patterned background.

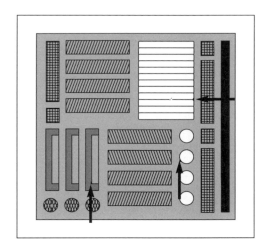

Answer to Figure 5.6.

Building quality into the product means controlling the manufacturing process and preventing mistakes. A process that meets the minimum standard for capability has a process capability index of 1.33. (The process capability index compares the process variation to the specification width. It measures the process' ability to meet the specification.) When such a process is under control, it produces 63.3 ppm (0.0063 percent) nonconforming products. A process with a capability index of 2 makes two bad pieces in every billion. We are unlikely to make a billion of anything, let alone assure this quality level by attribute inspection. Attribute data, the type of data that comes from inspections, cannot assure extremely high quality levels. We can, however, assure it by using quantitative (variables) data.

Nested Sources of Variation

SPC charts call for subgroups or samples of n measurements. We must be careful to select rational subgroups which represent homogeneous sets of process conditions. Batch processes, especially those in the microelectronics industry, complicate selection of the rational subgroup.

Figures 5.7 and 5.8 show the difference between sequential and batch processes. Figure 5.7 shows a process for coating semiconductor wafers with photoresist, and a drill press. In sequential processes, each piece is an independent representation of the process. In batch processes, each batch is an independent representation of the process. The pieces in a batch are not independent, since they share characteristics that depend on the batch. Consider an oven that bakes loaves of bread in batches of a dozen. We expect the loaves in one oven load to be like each other, but they may not be like those from another oven load. The variation between the loaves from the same oven load is *within-batch variation.* The variation between oven loads is *between-batch variation.*

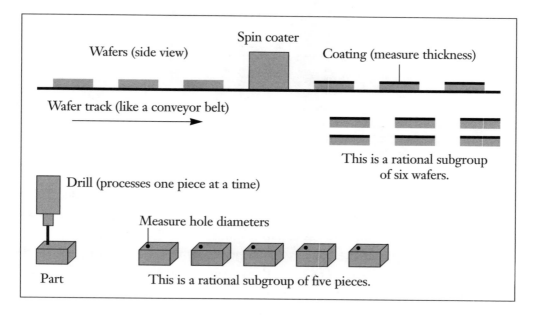

Figure 5.7. Sequential (conveyor belt) processing.

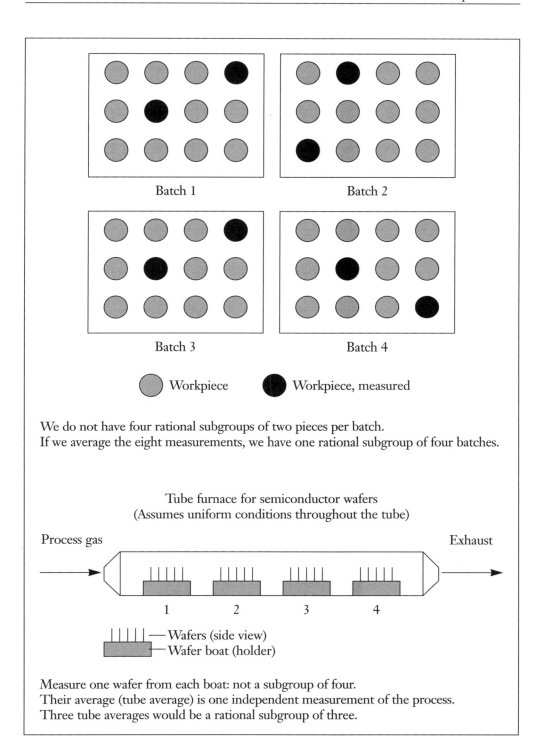

We do not have four rational subgroups of two pieces per batch.
If we average the eight measurements, we have one rational subgroup of four batches.

Measure one wafer from each boat: not a subgroup of four.
Their average (tube average) is one independent measurement of the process.
Three tube averages would be a rational subgroup of three.

Figure 5.8. Batch processing.

For example, it is a common mistake to take five pieces from a batch and call them a sample of five. The five pieces are not five independent representations of the process. The batch is one independent representation, and five batches would make a rational subgroup of five.

How can selecting the wrong subgroup cause trouble? The control limits for the chart for process average (X or $\bar{x}$ chart) depend on the process variation. Figure 5.9 shows how process variation for each piece includes both within- and between-batch variation. If we call samples from each batch rational subgroups, we will account only for the within-batch variation.

The process variation estimate will be too low, and the control limits will be too tight. The control chart for process mean will look like Figure 5.10. It's an immediate clue that this error is responsible. There are points near or over both the LCL and UCL. A point above the UCL suggests that the process mean has increased. A point below the LCL suggests that it decreased. Points outside both control limits suggest that the control limits are wrong! Figure 5.11 shows the same chart with the correct limits.

The chart for process variation (R chart or s chart) will look normal, but it will reflect the within-batch variation or batch uniformity. If the change involves between-batch variation, it will not tell us if the overall process variation changes.

This section will not go into the mathematical details of batch processes. The key point is that a chart that looks like Figure 5.10 suggests failure to account for all variation sources.

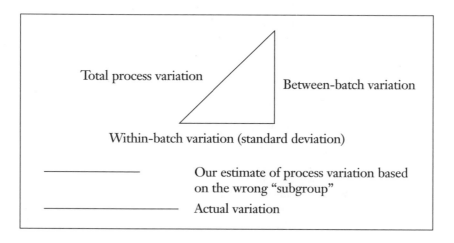

Figure 5.9. Process variation for batch processes.

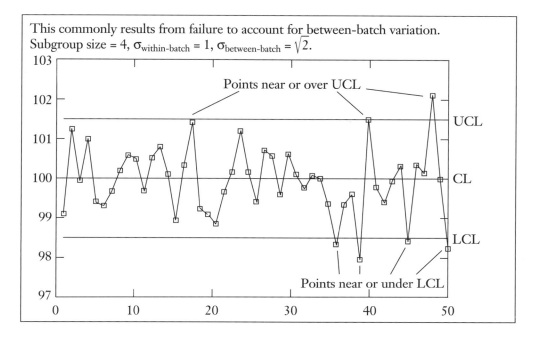

Figure 5.10. Control chart for process mean with incorrect control limits.

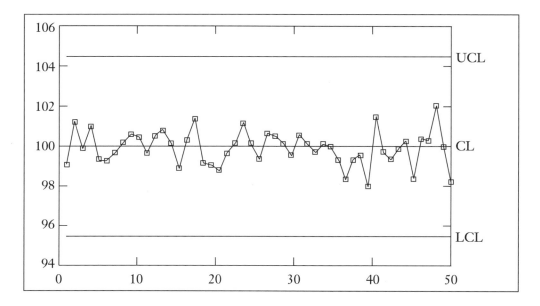

Figure 5.11. Same chart with correct control limits.

Multivariate Systems

When there are systematic within-batch differences, the system can be multivariate. This is also true when we measure two or more interdependent product characteristics. The mathematics are beyond the scope of this book, but we need to be aware of these situations.

Recall the bakery example. Suppose there are systematic temperature differences inside the oven. There may be hot and cold spots, so the product's condition will depend on its position in the oven. Table 5.3 shows, however, a correlation between all the loaves in the oven. This is a multivariate system.

Also consider the tube furnace for processing semiconductor wafers. To discuss multiple variation sources, we assumed that all the wafers in the tube receive equal process conditions. This is actually a poor assumption. As the process gases flow across the wafers, their composition changes. The wafers near the gas inlet use up some gas, so the wafers near the exhaust don't get as much gas. One such process uses steam and oxygen to grow a layer of silicon dioxide, which is chemically similar to glass, on the wafers. The layers on the wafers near the gas inlet are usually thicker than those near the exhaust. Figure 5.12 shows the correlation between the positions. If the layers near the gas inlet are thicker than usual, those at the other end also will be thicker than usual. The correct statistical model for this is a multivariate model.

Systematic differences across a process tool are usually undesirable, but are sometimes unavoidable. In the baking example, an improvement would force air to circulate in the oven. The air circulation, or convection, would help make the temperature uniform throughout the oven.

There also may be a correlation between product characteristics. This is common with products like transistors and rectifiers. Holmes (1988) and Montgomery (1991) discuss special control charts for these situations.

Table 5.3. Correlation.

	Condition of product 1 = very undercooked, 2 = undercooked, 3 = good, 4 = overcooked, 5 = burned		
	Oven load 1	Oven load 2	Oven load 3
Hot spot	5	4	3
Average position	4	3	2
Cold spot	3	2	1

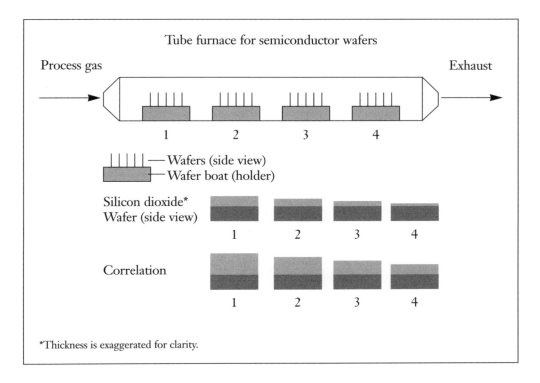

Figure 5.12. Silicon dioxide growth process for wafers.

Technical Appendix: Variance Components

This section discusses the mathematics of gage studies and nested variation sources. It shows how to quantify within-batch and between-batch variation and reproducibility and repeatability.

The expressions in Table 5.4 define the variances of gages and batch processes. Figure 5.13 shows them graphically. The standard deviation is the square root of the variance. The next sections show how to estimate these variance components.

Variance Components for Batch Processes

Suppose we have b batches and r_i measurements (replicates), and x_{ij}, in the ith batch. One-way analysis of variance (ANOVA) shows whether there is significant between-batch variation,* and it quantifies the variation. *One way* means there is one set of possible treatments—the batches. Table 5.5 shows a simple procedure for two nested sources of variation. For three or more sources, we must use techniques for nested experimental designs.

---

*Holmes and Mergen (1989) show how to use mean square successive differences (MSSDs) to test subgroups for rationality.

Table 5.4. Variance components for gages and batch processes.

Gage	Variance (σ^2_{gage}) of one measurement from one specimen	$\sigma^2_M + \sigma^2_R$	Variance of n measurements from one specimen	$\sigma^2_M + \dfrac{\sigma^2_R}{n}$
	M = Reproducibility, R = Repeatability			
Batch process	Variance ($\sigma^2_{\text{process}}$) of one piece from a batch	$\sigma^2_B + \sigma^2_W$	Variance of a sample of n pieces from a batch	$\sigma^2_B + \dfrac{\sigma^2_W}{n}$
	B = Between batches, W = Within batches			

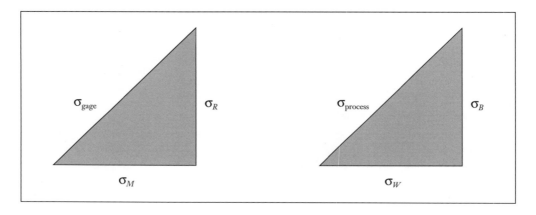

Figure 5.13. Variance components for gages and batch processes.

Example

A metallization chamber deposits metal on silicon wafers. The metallization tool is a vacuum chamber that evaporates or sputters a metal slug. The metal atoms condense on the wafers, which rest on rotating holders or planets. There is no systematic difference between the planets (or we would have a multivariate system), but there may be variation between runs. We measure one wafer from each of the three planets for each run, with the following results. The data come from a random normal simulator, with a mean of 1000 nanometers (nm), σ_{within} = 10 nm, and σ_{between} = 18 nm. Equation set 5.1 is for a system with two nested variation sources. $y \sim N(\mu, \sigma^2)$ means that y is a random number from a normal distribution with mean μ and variance σ^2. Here, y comes from variation between runs, and e reflects the within-run variation.

Table 5.5. Procedure for one-way ANOVA and variance components.

Step	Description	Formula
1	Find the sum of squares for errors (SSE). This will provide an estimate of the variation within batches.	$SSE = \sum_{i=1}^{b}\sum_{j=1}^{r_i}(x_{ij} - \bar{x}_i)^2$ Take the difference between each measurement and its batch average. Add the squares of these differences.
2	Calculate the sum of squares for treatments (SST). This will provide an estimate of the variance within and between batches.	$SST = \sum_{i=1}^{b} r_i(\bar{x}_i - \bar{\bar{x}})^2$
3	Calculate the mean square error (MSE) and mean square for treatments (MST). N is the total number of measurements, $= \sum r_i$.	$MSE = \dfrac{SSE}{N-b} \quad MST = \dfrac{SST}{b-1}$
4	Calculate the F statistic, and compare it to $F_{b-1;N-b;\alpha}$ where α is the desired significance level, or risk of concluding that there is batch-to-batch variation when there isn't. (Five percent is a typical significance level.) If $F > F_{b-1;N-b;\alpha}$, we are $100(1-\alpha)$ percent sure there is variation between the batches.	$F = \dfrac{MST}{MSE}$
5	Calculate the variance components. If all replicates (r_i) are equal.	$\sigma_W^2 = MSE \quad \sigma_B^2 = \dfrac{MST - MSE}{r}$
	If the r_i's are not equal, calculate as shown (Montgomery 1984, 71–74).	$\sigma_W^2 = MSE \quad \sigma_B^2 = \dfrac{MST - MSE}{r_0}$ where $r_0 = \dfrac{1}{b-1}\left[N - \dfrac{\sum_{i=1}^{b} r_i^2}{N}\right]$

(Eq. set 5.1)

	Equation	This example
Mean for the ith run	$\mu_i = \mu_{process} + y_i \sim N(0, \sigma^2_{between\ runs})$	$\mu_i = 1000 + y_i \sim N(0, 18^2)$
jth piece in ith run	$x_{ij} = \mu_i + e_{ij} \sim N(0, \sigma^2_{within\ run})$	$x_{ij} = \mu_i + e_{ij} \sim N(0, 10^2)$

Metallization data

Run	Metal thickness in nanometers			
	Planet A	Planet B	Planet C	Average
1	988.6	976.5	983.3	982.8
2	1018.4	1014.8	1006.8	1013.3
3	1063.8	1043.0	1047.3	1051.4
4	1005.5	977.3	1015.4	999.4
5	1024.2	1015.4	1019.9	1019.8
6	994.9	989.5	988.4	990.9
7	1007.2	1019.1	1008.4	1011.6
8	996.0	1005.1	1015.2	1005.4
9	1037.8	1032.9	1019.5	1030.1
10	1010.5	999.3	986.3	998.7
11	998.3	994.9	985.8	993.0
12	999.3	992.9	987.7	993.3
13	1033.6	1023.9	1023.8	1027.1
14	982.9	974.0	983.0	980.0
15	1029.7	1020.0	1036.7	1028.8
16	1032.6	1033.9	1018.4	1028.3
17	986	975.9	993.7	985.2
18	999.1	991.2	978.8	989.7
19	1022.5	1002.8	1011.9	1012.4
20	1006.8	1009.3	1001.1	1005.7
21	1020.3	1008.9	1003.2	1010.8
22	997.6	1004	1011.3	1004.3
23	983.8	961.4	999.4	981.5
24	968.9	981.2	983.9	978.0
25	1018.4	991.1	1017.4	1009.0

Microsoft Excel's data analysis package contains a one-way ANOVA tool that does most of the calculations automatically. Here are the results. Brackets contain explanatory comments. For example, 1048.475 is the MST, and 94.207 is the MSE.

ANOVA						
Source of variation	*SS*	*df*	*MS*	*F*	*P-value*	*F crit*
Between groups	25163.39 [SST]	24	1048.475 [MST]	11.12947	8.68E – 13	1.737078
Within groups	4710.353 [SSE]	50	94.20707 [MSE]			
Total	29873.75	74				

Since $F = 11.13$ is greater than $F_{24;50;0.05}$, we are more than 95 percent sure there is variation between runs or batches. *df* stands for degrees of freedom. Excel provides the P value, or α for $F = 11.13$. The chance that there is no between-run variation and that $F = 11.13$ is due only to chance, is 8.68×10^{-13}.

We must compute the variance components manually.

$$\sigma_{within} = \sqrt{MSE} = 9.71 \text{ (the simulation used 10)}$$

$$\sigma_{between} = \sqrt{\frac{MST - MSE}{r}} = \sqrt{\frac{1048.5 - 94.2}{3}} = 17.84 \text{ (the simulation used 18)}$$

$$\sigma_{process} = \sqrt{9.71^2 + 17.84^2} = 20.31$$

$$\sigma_{sample\ average,\ n\ =\ 3} = \sqrt{17.84^2 + \frac{9.71^2}{3}} = 18.70$$

If we take the standard deviation of the 25 averages, we get 18.69. This is very close to 18.70. The grand average of the 75 measurements is 1005.2. The correct control limits for an $\bar{x}$ chart are $1005.2 \pm 3 \times 18.7$. An incorrect chart would fail to account for the variation between runs and would use $1005.2 \pm 3 \times 9.71/3^{0.5}$. We would notice this very quickly, since many points would be outside both control limits.

Manugistics' StatGraphics 2.0 for Windows produces the following results from the same data. The software can handle several levels of nesting. StatGraphics 2.0 results for nested variation example areas follows:

Variance components analysis
Dependent variable: Thickness
Factors:
    Run

Number of complete cases: 75

Analysis of variance for thickness

Source	Sum of squares	df	Mean square	Var. comp.	Percent
TOTAL (CORRECTED)	29873.7	74			
Run	25163.4	24	1048.47	318.089	77.15
ERROR	4710.35	50	94.2071	94.2071	22.85

The StatAdvisor

The analysis of variance table shown here divides the variance of thickness into one component, one for each factor. Each factor after the first is nested in the one above. The goal of such an analysis is usually to estimate the amount of variability contributed by each of the factors, called the variance components. In this case, the factor contributing the most variance is run. Its contribution represents 77.1506 percent of the total variation in thickness.

The square root of the run variance is 17.84, and that of the error (within run) variance is 9.71. These match the results from the previous analysis.

Finally, consider the effect on process capability indices. The process capability indices reflect the process' ability to meet the specifications. Individual pieces (not samples) are in or out of specification. Therefore, we use $\sigma_{process}$ as the divisor for the indices.

Gage Studies

Hradesky (1988), Montgomery (1991), and Barrentine (1991) describe slightly different ways of analyzing gage studies. We will cover Barrentine's

(Eq. set 5.1 for gages)

	Equation	This example
Bias for the jth operator	$\tau_j \sim N(0, \sigma_M^2)$	$\tau_j \sim N(0, 3^2)$
kth piece measurement by jth operator, ith specimen	$x_{ijk} = d_i + \tau_j + e_{ijk} \sim N(0, \sigma_R^2)$	$x_{ijk} = d_i + \tau_j + e_{ijk} \sim N(0, 4^2)$

procedure, which uses the General Motors (GM) long form. The GM procedure uses a 5.15σ interval, which contains 99 percent of a normal distribution, to define R&R. Other references use a 6σ interval, which contains 99.73 percent of a normal distribution. Barrentine says to have at least two operators measure 10 or more specimens at least twice.

Example

The following simulation assumes a gage has the following R&R components: $\sigma_R = 4$ mils and $\sigma_M = 3$ mils. The equation set for nested models also applies to gages as follows: d_i is the true dimension of the specimen; τ_j is the bias for operator j; and e_{ik} is the random repeatability error for each measurement.

The model for measurement x_{ijk} (ith specimen, jth operator, kth replicate) is $x_{ijk} + d_i + \tau_j \sim N(0, \sigma_M^2) + e_{ijk} \sim N(0, \sigma_R^2)$. Four operators measure 10 specimens three times each with the following results. Assume that the specification for the item is [75, 125] mils. "Rep." stands for replicate or measurement repetition.

"Avg." is the average of the 30 measurements, and "Avg. R" is the average of the 10 ranges. The ranges provide information on the repeatability, or the gage's ability to return the same measurement for a specimen.

The range of the averages provides information on the reproducibility, or the ability of different operators to get the same measurement from a given specimen. We can compute the grand average range and the range between averages. Table 5.6 shows the R&R calculations.

	Operator 1				Operator 2			
Specimen	Rep. 1	Rep. 2	Rep. 3	R	Rep. 1	Rep. 2	Rep. 3	R
1	98.45	105.61	100.07	7.16	98.97	99.42	97.43	1.99
2	106.17	112.25	108.60	6.08	98.96	105.66	103.61	6.7
3	102.63	105.01	99.62	5.39	97.67	90.55	99.06	8.51
4	107.04	97.10	103.59	9.94	96.38	101.57	89.32	12.25
5	105.58	100.29	109.25	8.96	97.98	101.29	94.71	6.58
6	108.15	112.76	104.54	8.22	101.71	108.47	101.43	7.04
7	99.27	102.75	101.91	3.48	94.52	93.02	89.40	5.12
8	101.10	107.61	105.49	6.51	98.69	93.97	88.14	10.55
9	107.75	107.26	111.97	4.71	108.78	105.81	106.65	2.97
10	101.31	105.26	103.89	3.95	93.78	98.42	99.04	5.26
	Average	104.74	Average R	6.44	Average	98.48	Average R	6.70
	Operator 3				Operator 4			
Specimen	Rep. 1	Rep. 2	Rep. 3	R	Rep. 1	Rep. 2	Rep. 3	R
1	92.02	99.73	88.08	11.65	94.56	100.97	99.56	6.41
2	98.09	91.14	103.50	12.36	98.82	103.49	99.87	4.67
3	101.04	102.48	96.72	5.76	105.82	99.70	102.55	6.12
4	97.05	96.77	96.61	0.44	103.67	97.82	95.29	8.38
5	102.91	101.09	98.18	4.73	97.61	96.73	97.07	0.88
6	103.54	100.18	97.83	5.71	106.05	106.16	102.64	3.52
7	96.45	90.90	93.34	5.55	95.75	97.03	99.74	3.99
8	98.79	96.63	98.77	2.16	89.78	99.65	95.77	9.87
9	103.40	98.24	99.61	5.16	97.38	108.8	103.65	11.42
10	95.58	104.96	93.15	11.81	108.36	98.63	104.65	9.73
	Average	97.89	Average R	6.53	Average	100.25	Average R	6.50

Operator	Average range	Average
1	6.44	104.74
2	6.70	98.48
3	6.53	97.89
4	6.50	100.25
	Grand average R = 6.54	Range = 104.74 − 97.89 = 6.85

Table 5.6. R&R calculations.

	Formula	Example	Simulator used
Repeatability	$\sigma_R = \dfrac{\overline{\overline{R}}}{d_{2,m}}$ for m replicates	$\sigma_R = \dfrac{6.54}{1.693} = 3.86$	4
Reproducibility	$\sigma_M = \sqrt{\left(\dfrac{R_{avgs}}{d_{2,r}} c_{4,r}\right)^2 - \dfrac{\sigma_R^2}{nr}}$ For r operators and n specimens	$\sigma_M = \sqrt{\left(\dfrac{6.85}{2.059} 0.9213\right)^2 - \dfrac{3.86^2}{10 \times 4}}$ $= 3.00$	3
Gage	$\sigma_{gage} = \sqrt{\sigma_R^2 + \sigma_M^2}$	$\sigma_{gage} = \sqrt{3.86^2 + 3.00^2} = 4.89$	(5)*
PTCC or P/T ratio	$PTCC = \dfrac{5.15\sigma_{gage}}{USL - LSL}$	$PTCC = \dfrac{5.15 \times 4.89}{125 - 75} = 0.504$ or 50.4%	
D_4, d_2, and c_4 are control chart factors that depend on the sample size (Appendix E).			

*Since the simulator used $\sigma_R = 4$ and $\sigma_M = 3$, $\sigma_{gage} = 5$. The 5 is a function of the R&R elements. The simulator did not actually use 5 in a separate simulation. Instead, we expect the gage study to yield an estimate of 5 for σ_{gage}.

A range chart (Figure 5.14) shows whether any of the ranges are unusual, and we set it up like a standard range chart. The UCL is $D_4\overline{\overline{R}}$, which is $2.574 \times 6.54 = 16.83$ here. We would consider any sample with a range greater than 16.83 unusual.

The results from StatGraphics using the average and range method are in Figure 5.15. Results do not match exactly because of rounding in the calculations. The software also offers an ANOVA method. The average/range method reproduces the GM long form results.

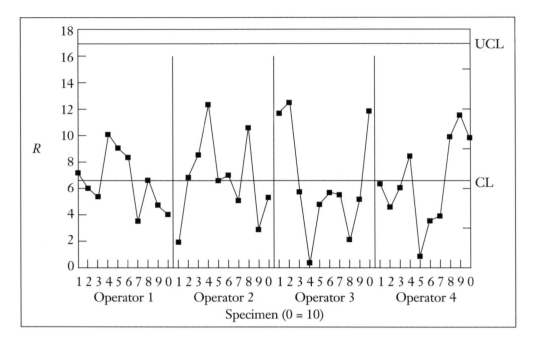

Figure 5.14. *R* chart for gage study.*

*Each section is really a separate range chart. Do not connect, for example, operator 1's tenth point and operator 2's first point.

We can suppress the repeatability variation by measuring each piece m times. The model for measurement x_{ijk} (ith specimen, jth operator, kth replicate) becomes

$$x_{ijk} = d_i + \tau_j \sim N(0, \sigma_M^2) + e_{ijk} \sim N\left(0, \frac{\sigma_R^2}{m}\right)$$

Then

$$\sigma_{\text{gage}} = \sqrt{\sigma_M^2 + \frac{\sigma_R^2}{m}}$$

The practicality of multiple measurements in a factory depends on the gage. For example, suppose that it takes two minutes to place the specimen in the gage (setup) and one second to measure it. Taking four measurements instead of one increases the time from 121 seconds to 124 seconds, or less than three percent. If, however, it takes 30 seconds to place the specimen in the gage and a minute to measure it, multiple measurements would be costly.

Analysis summary

Operator variable: Operator
Part variable: Specimen
Trial variable: Trial
Measurement variable: mils

Four operators   10 parts   Three trials

Average range = 6.54225   Range of *x*-bars = 6.85

	Estimated sigma	Estimated variance	Percent of total
Repeatability	3.8511	14.8309	62.61
Reproducibility	2.97611	8.85722	37.39
R&R	4.86705	23.6882	100.00

The StatAdvisor

Based on a study involving four operators, each measuring 10 parts three times, the estimated standard deviation of the measurement process for mils equals 4.86705. Of the total variance, 37.3909 percent is due to differences between operators (reproducibility) while 62.6091 percent is due to the instrument (repeatability). To estimate the range of error this implies, select Gage Report from the list of Tabular Options.

Figure 5.15. StatGraphics 2.0 results for gage study.

Gage Capability and Process Capability

Process capability estimates, discussed in chapter 6, come from measurements of the process. The variation of these measurements depends on the process variation and the gage variation, as follows: $\sigma_{measurement} = \sqrt{\sigma^2_{process} + \sigma^2_{gage}}$. The process capability supposedly reflects the process' ability to meet the specification. That is, we want $C_p = \dfrac{USL - LSL}{6\sigma_{process}}$ and not $C_p = \dfrac{USL - LSL}{6\sigma_{measurement}}$. Estimating the standard deviations from the measurements, however, gives us $\sigma_{measurement}$. We want to use this for control charts, since measurements are in or out of control. Parts, however, are in or out of specification, and $\sigma_{measurement}$ makes the process capability estimate too low. This effect is most noticeable when the process is very good and the gage is very bad.

Barrentine (1991, 41–43) defines the actual, or inherent, process capability as follows: C_{po} is the observed process capability estimate; that is, (USL – LSL)/($6\sigma_{measurement}$).

$$C_{pA} = \frac{1}{6\sqrt{\left(\frac{1}{6C_{po}}\right)^2 - \sigma_{gage}^2}} = \frac{1}{6\sqrt{\left(\frac{1}{6C_{po}}\right)^2 - \left(\frac{PTCC}{5.15}\right)^2}}$$

(Eq. 5.2)

Alternately, $C_{pA} = \dfrac{1}{\sqrt{\left(\frac{1}{C_{po}}\right)^2 - 36\left(\frac{PTCC}{5.15}\right)^2}}$

Figure 5.16 plots this equation for PTCC = 0 percent, 10 percent, 25 percent, and 30 percent. For example, if C_{po} is 1.4 and PTCC is 30 percent, the actual process capability is 1.605.

Figure 5.16 shows that a mediocre gage can significantly reduce the process capability estimate for a good process. When the process capability is poor, however, even a 30 percent PTCC has little effect on the estimate.

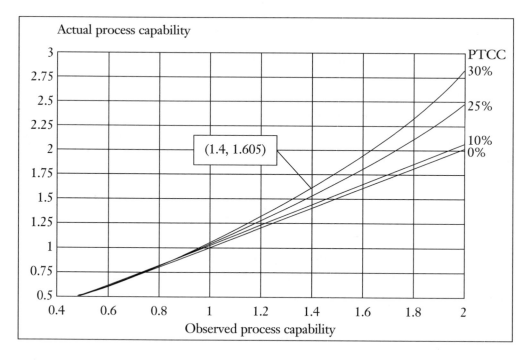

Figure 5.16. Inherent or actual process capability.

Remember that a noncapable process generates many nonconformances. A bad gage will allow some to pass, so we must replace the gage or improve the process. When the process capability is good, there will be few nonconformances (< 2 parts per billion when C_{pk} = 2). If such a process is in control, it is not urgent to replace the gage.

Gage Capability and Outgoing Quality

The contour plot in Figure 5.17 shows the effect of gage capability on outgoing quality. The figure uses a very incapable process and a poor gage to make all the contours large enough to see. It uses a process with specifications [2, 8], nominal = 5, σ_p = 1.5, and σ_{gage} = 0.5825. Each contour represents a constant probability density, with the highest in the center and the smallest

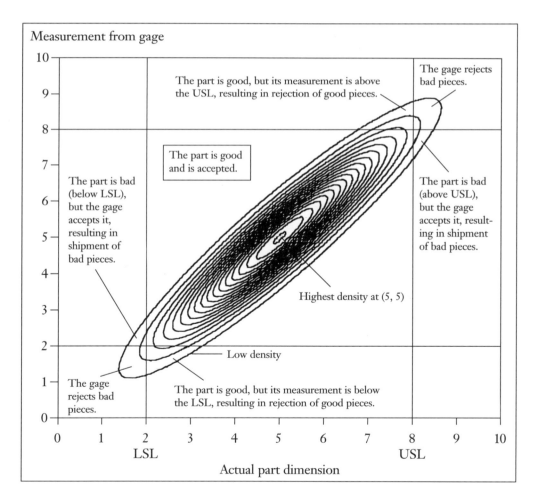

Figure 5.17. Gage capability and outgoing quality (C_p = 2/3, PTCC = 50 percent).

on the outside. Figure 5.17 illustrates the chances of passing bad pieces and rejecting good ones. Equation 5.3 shows the joint probability of making a piece with dimension x and getting measurement y.

$$f(x, y) = \frac{1}{2\pi\sigma_p\sigma_g} \exp\left[-\frac{1}{2\pi}\left(\left(\frac{x - \mu_p}{\sigma_p}\right)^2 + \left(\frac{y - x}{\sigma_g}\right)^2\right)\right] \qquad \textbf{(Eq. 5.3)}$$

In the equation, the process standard deviation is σ_p, that for the gage is σ_g, and the process mean is μ_p. Figure 5.17 shows the contour plot of Equation 5.3. Double integration of this expression yields the shipment of bad products, rejection of good units, and so on (Levinson 1996).

CHAPTER SIX

Process Characterization and Advanced Techniques

Chapter 3 showed how to understand and interpret control charts, but not how to set them up. Introducing a control chart requires *process characterization*, which requires us to estimate the process' mean and standard deviation. We must also test the assumption that the process follows a normal distribution. Some processes, especially those with one-sided specifications, follow other distributions.

The Normal Distribution

The normal distribution is a good model for most manufacturing processes. We can use it to calculate the process yield and the chance of getting a point outside the control limits.

A statistical distribution's *probability density function* (pdf) shows the relative chance of getting a number from a population. For the *discrete distributions* (binomial and Poisson), this chance is an actual percentage. Suppose a process' rejection rate is 2 percent. We have a 13.3 percent chance of getting exactly zero bad pieces from a sample of 100. The normal distribution is, however, a *continuous distribution*. There is a differential or very tiny chance of getting a particular number. Thus, we must compute the chance of getting a range of numbers. For example, suppose the mean is 5.0000 and the standard deviation is 0.5. There is no chance of getting exactly 5.0000, although this is the most likely result. There is, however, a 68.27 percent chance of getting a number between 4.5 and 5.5.

Why should this calculation interest us? In the example, suppose that 4.5 and 5.5 are the specification limits. This would not be a very good process, since only 68.27 percent of the product would be good. If the specifications were [2, 8], however, we'd expect only two bad pieces in every billion. This would be a very good process. No matter what the specification limits are, the control limits for a single measurement are [3.5, 6.5]. If the process is in control, there is only a 0.27 percent chance that a measurement will be outside this range. These chances—68.27 percent, two parts per billion, 0.27 percent—all come from the cumulative normal distribution, which we will discuss shortly.

Two parameters define a normal distribution. The *mean*, or *location parameter*, defines the center of gravity. The standard deviation, or shape parameter, defines the variation or spread. The Greek letter mu (μ) symbolizes the mean, and the Greek letter sigma (σ) symbolizes the standard deviation.

It is conventional to use Greek letters to symbolize actual or true values. English letters refer to values from samples. A Greek letter with a caret ($\wedge$) or "hat" over it is an estimate.

Parameter	Population	Estimate	Sample
Mean	μ	$\hat{\mu}$	$\bar{x}$
Standard deviation	σ	$\hat{\sigma}$	s

Equation 6.1 shows the probability density function for the normal distribution. It defines the relative chance of getting value x from a normal distribution with mean μ and standard deviation σ. The shorthand form for "x comes from a normal distribution with mean μ and variance σ^2" is $x \sim N(\mu, \sigma^2)$.

$$f(x) = \frac{1}{\sigma\sqrt{2\pi}} \exp\left(-\frac{1}{2}\left(\frac{x - \mu}{\sigma}\right)^2\right)$$ (Eq. 6.1)

Figure 6.1 shows a graph of this function for $x \sim N(10, 0.5^2)$, $x \sim N(10, 1^2)$, and $x \sim N(10, 2^2)$. Note that

- The mean 10 is the most likely value or *mode.*

- The curves are symmetric around the mean. Half of the area under the curve is on either side of the mean. This makes the mean equal to the median.

  —A probability density function's median is its 50th percentile. Half of the distribution is on either side of the median.

- The standard deviation defines the curve's shape. A high standard deviation results in a broad function, while the curve for a low standard deviation is narrow.

Cumulative Normal Distribution

The cumulative normal distribution lets us find the chance of getting a value within a range of numbers. This range could be the specification or the control limits.

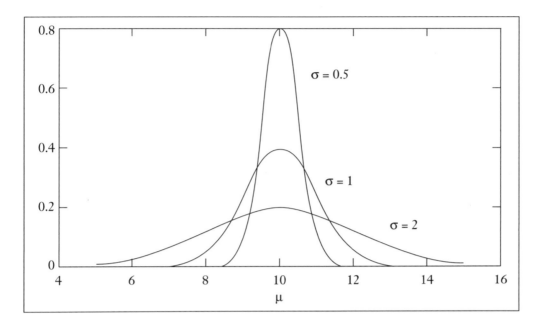

Figure 6.1. Normal probability density function.

The normal probability density function itself has little use in industrial statistics.* This is because there is an infinitesimal chance of getting any specific value from a population. Our true concern is the chance of getting a value that is within, or outside, a range. To calculate this, we must integrate the pdf. Here is the chance of getting a value from the range [*a*, *b*].

$$\Pr(a \leq x \leq b) \;=\; \int_a^b \frac{1}{\sigma\sqrt{2\pi}} \exp\!\left(-\frac{1}{2}\left(\frac{y-\mu}{\sigma}\right)^2\right)dy$$

$\Pr(a \leq x \leq b)$ is shorthand for "probability that *x* is in the range [*a*, *b*]." $\int$ is an integration sign. $\int_a^b f(y)dy$ is the area under function *f* in the range [*a*, *b*].

We can simplify this expression by using the *standard normal deviate z*. This standardizes a value by expressing it in standard deviations from the mean. The value is *z* standard deviations from the population mean. Then

---

*In this book, its sole application is to generate bell curve figures. Applications rely on the cumulative normal distribution or integrals of the pdf.

we can use standard tables of, or software functions for, the *cumulative standard normal distribution* $\Phi(z)$.

$$z = \frac{x - \mu}{\sigma} \text{ or } x = \mu + z\sigma, \qquad \textbf{(Eq. 6.2)}$$

where x is z standard deviations from the mean.

$$\Phi(z) = \int_{-\infty}^{z} \frac{1}{\sqrt{2\pi}} \exp\left(-\frac{1}{2} y^2\right) dy \qquad \textbf{(Eq. 6.3)}$$

Also, $\Phi(-z) = 1 - \Phi(z)$

We cannot integrate Equation 6.3 to get a closed form solution, but spreadsheets have functions for it. In Microsoft Excel 5.0 and Corel* Quattro Pro, NORMSDIST returns $\Phi(z)$, while the Lotus 1-2-3 function is NORMAL. MathCAD's Cumulative NORMal (cnorm) function returns $\Phi(z)$. Appendix A tabulates $\Phi(z)$ for standard normal deviates ranging from 0 to 4. Most statistics books have similar tables.

What happens if z is negative? The table shows values only for positive standard normal deviates. There are two important characteristics of the normal distribution.

1. The area under the pdf is 1. This applies to all continuous and discrete probability functions. The function must account for 100 percent of the population. Integration, or summation for discrete functions, over the population's range must yield 1.

2. The normal distribution is symmetric, so its lower half is the mirror image of its upper half. The standard normal distribution's mean is 0, so its negative side mirrors its positive side. For example, the section below –1 is the mirror image of the section above 1. The area above $z = 1$ equals $1 - \Phi(1)$, since $\Phi(1)$ is the integral from $-\infty$ to 1. Figure 6.2 shows that the area below $z = -1$ is its mirror image, so $\Phi(-1)$ equals $1 - \Phi(1)$. As Equation 6.3 shows, $\Phi(-z) = 1 - \Phi(z)$.

How can we use this to calculate the fraction of a population that is between a and b? Suppose a drill press makes holes whose nominal diameter is 1/8 inch (125 mils). The specification is [123, 127] mils, the process mean is 124.5 mils, and the standard deviation is 1 mil. Here is a summary of the calculations, and Figure 6.3 illustrates them. Appendix A contains a table for $\Phi(z)$.

---

*Borland, and then Novell, previously sold Quattro Pro.

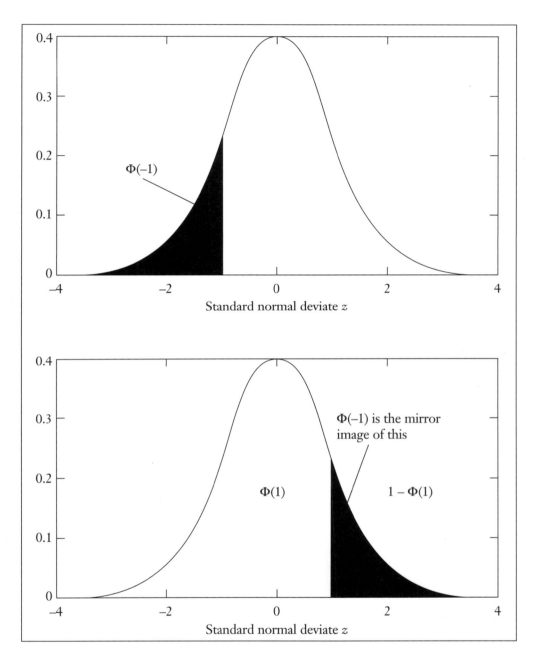

Figure 6.2. Cumulative standard normal distribution.

Portion	Standard normal deviate	$\Phi(z)$
Below the USL	$z = \dfrac{(127 - 124.5)\text{ mils}}{1\text{ mil}} = 2.5$	$\Phi(2.5) = 0.993790$ $(1 - 0.99379 = 0.00621$ is above the USL.)
Below the LSL	$z = \dfrac{(123 - 124.5)\text{ mils}}{1\text{ mil}} = -1.5$	$\Phi(-1.5) = 1 - \Phi(1.5)$ $= 1 - 0.93319$ $= 0.06681$
Summary	0.99379 (below USL) − 0.06681 (below LSL) = 0.92698 92.698 percent of the product is in specification. 6.681 percent is below the LSL. 0.621 percent is above the USL.	

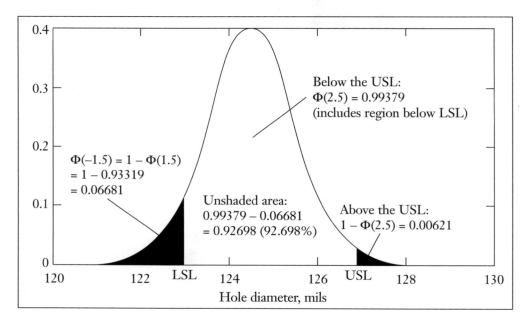

Figure 6.3. Application of the cumulative normal distribution.

This process is 0.5 mil off nominal, and we could improve it by increasing the process mean to 125.0 mils. Figure 6.4 shows the yield if we do this.

Finally, what is the chance of getting a point outside the Shewhart three-sigma control limits if the process is in control? If the process is in control, each control limit is three standard deviations from the mean. The chance of being below the UCL is $\Phi(3) = 0.998650$, so the chance of being over the UCL is 0.00135 (0.135 percent). The chance of being below the LCL is $\Phi(-3) = 1 - \Phi(3) = 0.00135$ (0.135 percent). Therefore, the chance of being below the LCL or above the UCL is 0.270 percent, or 2.7 chances per 1000 samples.

Portion	Standard normal deviate	$\Phi(z)$
Below the USL	$z = \dfrac{(127 - 125)\,\text{mils}}{1\,\text{mil}} = 2.0$	$\Phi(2.0) = 0.97725$ $(1 - 0.97725 = 0.02275$ is above the USL.)
Below the LSL	$z = \dfrac{(123 - 125)\,\text{mils}}{1\,\text{mil}} = -2.0$	$\Phi(-2.0) = 1 - \Phi(2.0)$ $= 1 - 0.97725$ $= 0.02275$
Summary	0.97725 (below USL) − 0.02275 (below LSL) = 0.95450 95.450 percent of the product is in specification. 2.275 percent is below the LSL. 2.275 percent is above the USL.	

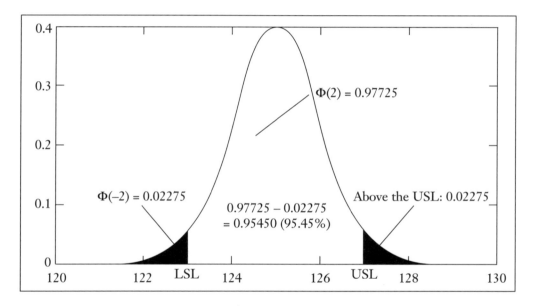

Figure 6.4. Application of the cumulative normal distribution (continued).

Tests for Normality

Standard methods for SPC, especially the Shewhart chart, rely on the assumption that the process follows a normal distribution. This section describes statistical tests for this assumption.

The chi square test is a quantitative assessment of how well a histogram fits a statistical distribution. It requires at least 30, and preferably 100 or more, data. The results depend on the user's selection of the histogram cells. The normal probability plot is a qualitative and quantitative tool for assessing normality.

Chi Square Test for Goodness of Fit

We examined the histogram, which shows frequency of occurrence. We said the histogram should have a bell shape if the population follows a normal distribution. The criterion, "looks like a bell," is subjective. The chi square test quantifies this assessment of a histogram. It returns a statistic, χ^2 (chi square), that measures the discrepancy between the histogram and the normal distribution. The test is also useful for checking data to see whether they conform to other distributions. For example, nonconformance (rework/scrap) data should follow the binomial distribution if the process is in control. Here is the procedure for the chi square test.

1. Prepare a histogram of the data.

   a. For n data, start with $\sqrt{n}$ cells (Messina 1987, 16), or $4(0.75(n-1)^2)^{0.2}$ cells (Shapiro 1986, 24). According to Juran and Gryna (1988, 23.72), n should be at least 30 and preferably 100 or more.

   b. Count the observations in each cell. Call this f_i (frequency in cell i) or O_i (observations in cell i).

2. Compute the expected number of observations in each cell.

   a. The null hypothesis (assumption) is that the data follow the normal distribution $N(\mu, \sigma^2)$.

      1. μ and σ^2 may be standards or givens.

      2. Alternately, $\bar{\bar{x}}$ and s^2 from the data are estimators for μ and σ^2.

   b. L_i is the lower limit for the ith cell, and U_i is the upper limit.

   c. If the population is normal, we expect E_i observations in the ith cell.
   $$E_i = n \int_{L_i}^{U_i} f(x)dx = n\left[\Phi\left(\frac{U_i - \mu}{\sigma}\right) - \Phi\left(\frac{L_i - \mu}{\sigma}\right)\right],$$
   where $f(x)$ is the normal pdf. Figure 6.5 illustrates this calculation for a population whose hypothetical distribution is $N(1000, 10^2)$ and cell [987.5, 993.75].

   d. The theory behind the chi square test relies on the assumption that E_i is 5 or greater. If this is not true, combine cells to meet this condition. This usually happens in the tail areas.

3. Compute the chi square test statistic. For k cells,

   a. $$\chi^2 = \Sigma_{i=1}^{k}\frac{(O_i - E_i)^2}{E_i}$$

   b. χ^2 is the sum of the evidence against normality (see Figure 6.6).

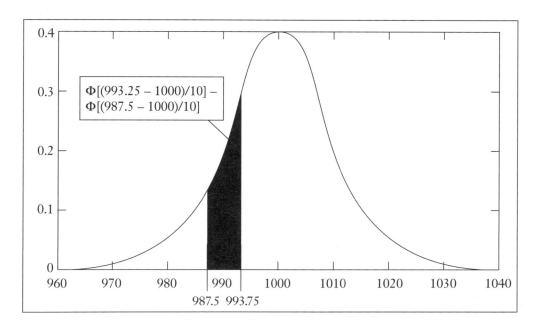

Figure 6.5. Expected cell count, chi square test.

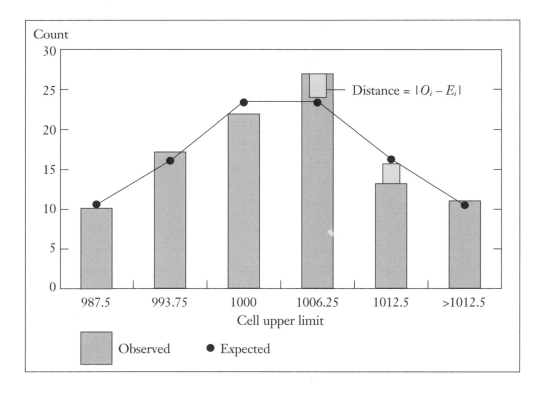

Figure 6.6. Chi square test and evidence against the normality assumption.

4. Reject the null hypothesis (normality assumption) if $\chi^2 > \chi^2_{k-1-p;\,\alpha}$.

   a. There are $k - 1 - p$ degrees of freedom when we estimate p parameters from the data.

      1. If we estimate μ and σ^2 from the data, then $p = 2$, and there are $k - 3$ degrees of freedom.

      2. If we use standards for μ and σ^2, $p = 0$, and there are $k - 1$ degrees of freedom.

   b. The significance level for the test is α, and is usually 0.05 (5 percent). It is the risk of wrongly concluding that a normal distribution is not normal.

      1. The test will, on average, reject the normality hypothesis five times out of 100 when the population is normal.

      2. See Appendix B for a chi square table; α is 1 minus the cumulative distribution. Suppose we want χ^2 for five degrees of freedom and $\alpha = 0.05$. We look up the 95th percentile of the distribution, which is 11.07. The upper tail (α) is $1 - 0.95 = 0.05$.

Example

One hundred measurements come from a population. We want to test the hypothesis that its mean is 1000 and its variance 100. The data fall into a histogram whose cell limits are as follows:

Upper limit	Observations
975	1
981.25	1
987.5	8
993.75	17
1000	22
1006.25	27
1012.5	13
1018.75	8
1025	2

We compute the expected count for each cell.

Upper limit	Observed	Expected	Calculation
975	1	0.62	$100\Phi\left(\dfrac{975 - 1000}{10}\right)$
981.25	1	2.42	$100\left(\Phi\left(\dfrac{981.25 - 1000}{10}\right) - \Phi\left(\dfrac{975 - 1000}{10}\right)\right)$
987.5	8	7.53	$100\left(\Phi\left(\dfrac{987.5 - 1000}{10}\right) - \Phi\left(\dfrac{981.25 - 1000}{10}\right)\right)$
993.75	17	16.03	
1000	22	23.40	
1006.25	27	23.40	
1012.5	13	16.03	
1018.75	8	7.53	
1025	2	2.42	
> 1025	1	0.62	$100\left(1 - \Phi\left(\dfrac{1025 - 1000}{10}\right)\right)$

We expect only 0.62 observations in the lowest cell and 2.42 in the second lowest. To get $E_i \geq 5$, we must combine the first three cells. Similarly, we must combine the three highest cells.

Lower limit	Observed	Expected
987.5	10	10.56
993.75	17	16.03
1000	22	23.40
1006.25	27	23.40
1012.5	13	16.03
≥1012.5	11	10.56

Finally, we perform the chi square calculation.

Lower limit	Observed	Expected	χ^2	Sample calculation
987.5	10	10.56	0.030	$\dfrac{(10 - 10.56)^2}{10.56}$
993.75	17	16.03	0.058	$\dfrac{(17 - 16.03)^2}{16.03}$
1000	22	23.40	0.084	
1006.25	27	23.40	0.553	
1012.5	13	16.03	0.574	
≥ 1012.5	11	10.56	0.018	$\dfrac{(11 - 10.56)^2}{10.56}$
Total			1.318	

There are six cells. Since we used standards for μ and σ^2, we estimated no parameters from the data. Compare 1.318 against $\chi^2_{5;\,0.05} = 11.07$. Since 1.318 is less, we accept the hypothesis that the population is normal. Remember that χ^2 measures the evidence against the normality assumption. We need 11.07 to reject the assumption with 95 percent confidence, but we have only 1.318.

Here is an example of a non-normal population. The data came from a bimodal simulation, $x = 85 + e \sim N(0, 10) + (30, 50$ percent chance$)$. That is, there is a 50 percent chance that the process mean is 85, and a 50 percent chance it is 115. The data are as follows:

104.3	126.6	116.9	89.9	87.4
127.1	123.2	90.1	86.8	79.1
100.9	111.0	122.2	122.8	123.4
75.5	93.2	95.2	118.3	108.5
130.8	91.9	78.8	126.5	83.1
82.9	78.9	98.9	112.5	113.7
70.4	109.3	108.2	119.0	108.6
125.6	83.6	94.0	101.0	88.5
91.0	66.6	75.8	112.0	88.5
110.5	88.5	116.2	83.1	118.8
108.5	116.7	115.9	81.3	79.6
123.9	112.5	121.5	117.1	106.3
136.1	87.2	120.2	105.7	87.0

77.7	88.4	63.4	76.4	124.7
79.4	86.8	73.8	103.5	129.0
117.2	131.6	78.9	113.1	117.3
126.4	92.1	85.8	104.6	109.7
121.8	81.3	123.2	83.4	113.0
102.2	112.4	66.7	82.1	80.7
77.3	83.5	88.4	68.1	122.3

It is easy to handle the test with a spreadsheet. Here is the solution with Microsoft Excel. The 100 data are in cells A5:E24. The MAX and MIN functions show the range of the data and allow the user to easily define a range of cells. AVERAGE returns $\bar{\bar{x}}$, and STDEV returns s ($\hat{\sigma}$).

Formula	Result
MAX(A\$5:E\$24)	136.1
MIN(A\$5:E\$24)	63.4
AVERAGE (A\$5:E\$24)	100.7
STDEV(A\$5:E\$24)	18.8

We start with the cell limits in the left column (step 1) (see Table 6.1). The TOOLS—DATA ANALYSIS menu has a HISTOGRAM function. The input range covers the data (A5:E24), and the function also asks for the cell limits. These are 65, 73, . . ., 137. The function returns the information under bin and frequency (step 2). We use the formula

$$E_i = n\left[\Phi\left(\frac{U_i - \mu}{\sigma}\right) - \Phi\left(\frac{L_i - \mu}{\sigma}\right)\right]$$ to compute the expected count in

each cell (step 3). Excel (and Corel Quattro Pro) function NORMSDIST(z) returns $\Phi(z)$. Lotus 1-2-3 uses NORMAL. We combine cells so the expected count is at least 5 (step 4). Then we compute $\dfrac{(O_i - E_i)^2}{E_i}$ for each cell, and add the results. To check steps 3 and 4, the observed frequency and expected counts must add to $n = 100$.

After combining the cells, there are nine. We estimated the mean and standard deviation from the data, so there are $9 - 3 = 6$ degrees of freedom. $\chi^2_{6;\,0.05} = 12.59$, which is less than 28.04, so we are 95 percent sure the population is not normal. Figure 6.7 shows the difference between the observed and expected counts.

The chi square test also works for other distributions, like the binomial (rework/scrap) and Poisson (defect). For the chi square test, these require constant sample sizes. For N samples of n, Equation set 6.4 shows the expected counts.

Table 6.1.

Step 1	Step 2		Step 3	Step 4		Step 5
Cell (upper limit)	Bin	Frequency	Expected	Frequency	Expected	Chi square
65	65	1	2.90			
73	73	4	4.17	5	7.07	0.61
81	81	13	7.73	13	7.73	3.60
89	89	20	11.97	20	11.97	5.38
97	97	8	15.52	8	15.52	3.65
105	105	7	16.84	7	16.84	5.75
113	113	15	15.29	15	15.29	0.01
121	121	13	11.61	13	11.61	0.17
129	129	15	7.38	15	7.38	7.87
137	137	4	3.92	4	6.59	1.02
	More	0	2.66			
Total		100	100	100	100	28.04

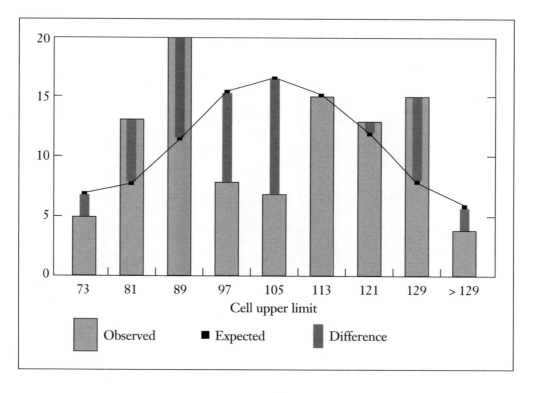

Figure 6.7. Chi square test and evidence against the normality assumption.

Eq. set 6.4. Expected counts for chi square test (summary).

		Degrees of freedom, k cells	
Distribution	Expected count, ith cell	Standard parameters	Empirical parameters
Normal	$E_i = n\left[\Phi\left(\dfrac{U_i - \mu}{\sigma}\right) - \Phi\left(\dfrac{L_i - \mu}{\sigma}\right)\right]$	$k - 1$	$k - 3$
Binomial	$E_i = N\dfrac{n!}{i!(n - i)!}p^i(1 - p)^{n - i}$ (N samples of n, i nonconformances)	$k - 1$	$k - 2$
Poisson	$E_i = N\dfrac{(np)^i}{i!}e^{-np} = N\dfrac{\mu^i}{i!}e^{-\mu}$ (N samples of n, i defects)	$k - 1$	$k - 2$

The Normal Probability Plot

The normal probability plot is another useful way to check the normality assumption. It requires less data than the chi square test, but we must remember that less data always reduce a test's power. A test that would detect non-normality with 50 data might not detect it with 10 data.

We order the data from smallest to largest; that is, $x_{(1)} \leq x_{(2)} \leq \ldots \leq x_{(n)}$. $x_{(i)}$ is the ith ordered or sorted datum. Spreadsheets and MathCAD have functions or tools that do this automatically.

Suppose there are 10 data. The lowest point represents the 0 to 10th percentile, and this range's midpoint is 5 percent. The second lowest point represents the 10th to 20th, and the midpoint is 15 percent. The highest point represents the 90th to 100th, and the midpoint is 95 percent. In general, we put the ith point at the $\dfrac{i - 0.5}{n}$ percentile.* That is, we assume $F(x_{(i)}) = \dfrac{i - 0.5}{n}$.

Now we use the equation for the standard normal deviate and rearrange it.

$$z = \frac{x - \mu}{\sigma} \Rightarrow x = \mu + z\sigma \Rightarrow x_{(i)} = \mu + z_{(i)}\sigma, \text{ where } \Phi(z_{(i)}) = \frac{i - 0.5}{n}$$

Therefore, a plot of $x_{(i)}$ versus the ordered standard normal deviates $z_{(i)}$ should be linear with slope σ and intercept μ.

It is not convenient to look up the $z_{(i)}$s manually, especially if there are a

---

*Other references use $(i - 0.3)/(n + 0.4)$, $(i - 0.375)/(n + \frac{1}{4})$, and $i/(n + 1)$.

lot of them. Normal probability paper is available, with a percentile scale that is linear in z. To use it, we must calculate $(i - \frac{1}{2})/n$ for each point. The process is easy with a spreadsheet or MathCAD. Excel and Quattro Pro's NORMSINV function returns z for a percentile of the normal distribution. Lotus uses the NORMAL function, which is also the function for $\Phi(z)$. An argument of this function tells it what to calculate. In MathCAD, root(cnorm(dummy) $- (i - \frac{1}{2})/n$, dummy) finds z such that $\Phi(z) - (i - \frac{1}{2})/n$ is zero. "Dummy" is a dummy variable, which requires an initial guess. Zero is a good initial guess. Figure 6.8 shows a normal probability plot of 50 data, $x \sim N(1000, 10^2)$.

The points fit the line well. The slope is 9.808, and the sample standard deviation is 9.791. The intercept and average are 1001.6. The intercept will always be close to the average, even if the distribution is not normal. The slope will often be close to the sample standard deviation, too. If the distribution is non-normal, however, the correlation will be poor. There may be a systematic pattern of points around the best fit line, when they should scatter randomly. The slope may not be close to σ. Figure 6.9 shows a probability plot for non-normal data.

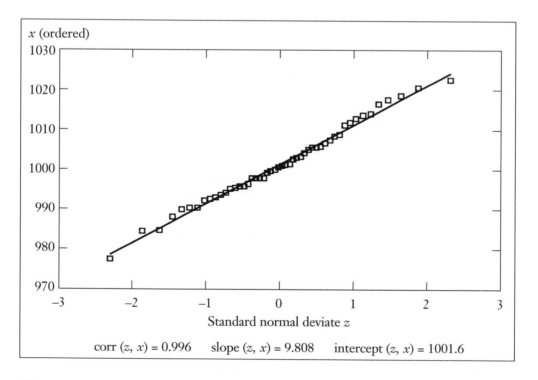

$$\text{corr}\ (z, x) = 0.996 \qquad \text{slope}\ (z, x) = 9.808 \qquad \text{intercept}\ (z, x) = 1001.6$$

Figure 6.8. Normal probability plot.

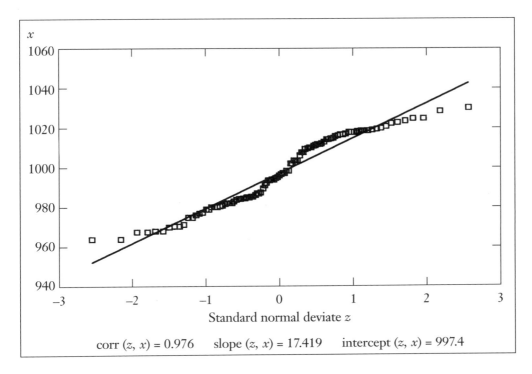

corr $(z, x) = 0.976$ slope $(z, x) = 17.419$ intercept $(z, x) = 997.4$

Figure 6.9. Normal probability plot, bimodal distribution (100 data).

Here, the correlation is not very good, and there is a systematic pattern to the points. They should scatter randomly around the best fit line, but they don't.

Not very good is a subjective statement. Cryer (1986, 43) gives a table of critical values for correlation in the normal probability plot. This information appears in Table 6.2. Cryer uses $(i - 0.375)/(n + \frac{1}{4})$ for the percentile of z instead of $(i - \frac{1}{2})/n$, but this should make little difference. We can reject the normality hypothesis with $100(1 - \alpha)$ percent confidence if the correlation is less than the value in the table. For example, 99 out of 100 samples of 100 from a normal population will, on average, yield correlations of 0.981 or better. Ninety percent of the samples will yield correlations of 0.989 or better. Here, $0.976 < 0.981$, so we are better than 99 percent sure that the sample did not come from a normal population.

Figure 6.9a shows a plot of bimodal data from StatGraphics 2.0. This is the same data that we used for the chi square test: $x = 85 + e \sim N(0, 10) + (30, 50$ percent chance). That is, there is a 50 percent chance that the process mean is 85, and a 50 percent chance it is 115. The axis is linear in z. The proportion is $\Phi(z)$.

Table 6.2. Critical values for correlation, normal probability plot.

	Significance level α		
Sample size	0.10	0.05	0.01
10	0.935	0.918	0.880
15	0.951	0.938	0.911
20	0.960	0.950	0.929
25	0.966	0.958	0.941
30	0.971	0.964	0.949
40	0.977	0.972	0.960
50	0.981	0.976	0.966
60	0.984	0.980	0.971
75	0.987	0.984	0.976
100	0.989	0.986	0.981
150	0.992	0.991	0.987
200	0.994	0.993	0.990

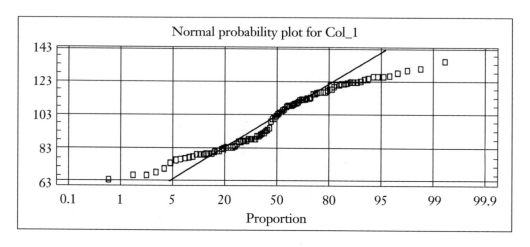

Figure 6.9a. Normal probability plot, bimodal distribution (StatGraphics 2.0).

If we have n samples, each with m measurements, we can do a probability plot of the sample variances. We plot the ordered sample variances, $s_{(i)}^2$, against the $(i - \frac{1}{2})/n$ percentile of the chi square distribution with $m - 1$ degrees of freedom. This requires a computer. Microsoft Excel and Corel Quattro Pro's CHIINV function, and Lotus' CHIDIST function, calculate

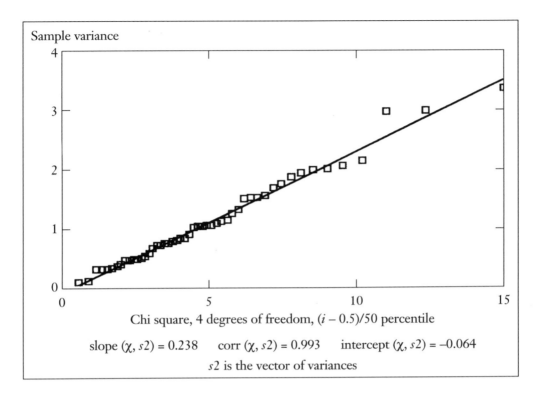

Figure 6.10. Chi square probability plot.

these percentiles. Alternately, we can use MathCAD's root function with the chi square probability density function. (This function appears in Appendix B.) The plot should be linear with slope $\sigma^2/(m-1)$ and intercept 0.

Figure 6.10 shows a chi square probability plot for 50 samples of five from a process whose standard deviation is 1. The points fit well, the slope is 0.238 ($\approx 1^2/(5-1) = 0.250$), and the intercept is close to 0.

Central Limit Theorem

Averages from large samples follow a normal distribution even when they come from non-normal populations. This helps us make control charts for non-normal populations.

One way to handle non-normal populations is to characterize the probability distribution. Lawless (1982) shows how to fit data to gamma, Weibull, lognormal, and other distributions. Setting the LCL at the 0.135 percentile and the UCL at the 99.835 percentile yields Shewhart-equivalent false alarm risks.

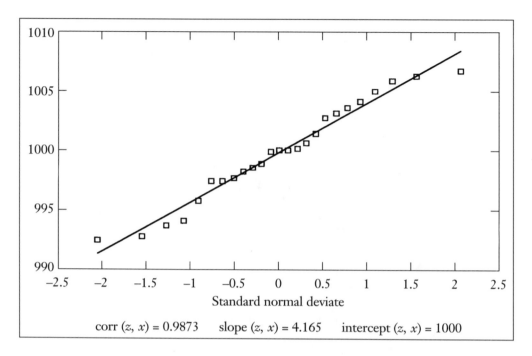

Figure 6.11. Normal probability plot, samples of 10 from bimodal distribution.

If the samples are large enough, however, this may be unnecessary. The central limit theorem says that averages of infinitely large samples follow the normal distribution. In practice, a sample of 5 or 10 may be adequate. It depends on how non-normal the population is. Figure 6.11 shows a normal probability plot for 25 samples of 10 from a bimodal distribution, where $x = 985 + e \sim N(0, 10) + (30, 50$ percent chance$)$. The points fit reasonably well, and the correlation meets the requirements in Table 6.2.

Setting up a Control Chart

After testing the normality assumption, we're ready to set control limits for a control chart. There are several ways to do this. *Theoretical charts* assume that we know the process mean and standard deviation. This applies to processes for which we've had long experience or processes for which there are a lot of data. *Empirical charts* rely on estimates for the mean and standard deviation.

Theoretical Control Charts

Equation set 6.5 defines the control limits for a sample of n measurements.

Eq. set 6.5. Control limits, given μ and σ.

	Chart for process mean	Chart for process variation	
Estimate for standard deviation	$\bar{x}$ chart	s (sample standard deviation) chart	R (sample range) chart
Given	$\mu \pm 3\dfrac{\sigma}{\sqrt{n}}$	Limits: $[B_5\sigma, B_6\sigma]$ Centerline: $c_4\sigma$	Limits: $[D_1\sigma, D_2\sigma]$ Centerline: $d_2\sigma$

B_5, B_6, D_1, D_2, d_2, and c_4 are control chart factors that depend on the sample size. Appendix E tabulates them for samples of up to 25. The s chart for samples of five or less, and the range chart for samples of six or less, have no LCLs.

Empirical Control Charts

Suppose we have m samples, each of size n_i. For each sample, we have its standard deviation s_i or range R_i. The standard deviation is better, but it is not practical for manual calculations. The grand average of all the data is $\bar{\bar{x}}$. Equation set 6.6 shows how to set the control limits.

Eq. set 6.6. Control limits, given estimates of μ and σ.

	Estimate for standard deviation	Chart for process mean	Chart for process variation	
		$\bar{x}$ chart	s chart	R (sample range) chart
Based on s	$\hat{\sigma} = \dfrac{1}{m}\Sigma_{i=1}^{m}\dfrac{s_i}{c_4(n_i)}$	$\bar{\bar{x}} \pm A_3\bar{s}$	$[B_3\bar{s}, B_4\bar{s}]$ Centerline: $\bar{s}$	
Based on ranges	$\hat{\sigma} = \dfrac{1}{m}\Sigma_{i=1}^{m}\dfrac{R_i}{d_2(n_i)}$	$\bar{\bar{x}} \pm A_2\bar{R}$		$[D_3\bar{R}, D_4\bar{R}]$ Centerline: $\bar{R}$
When the samples sizes vary, so do the control limits and even the centerlines of the s and R charts. For a sample of k, $\bar{s}(k) = c_{4,k}\hat{\sigma}$ and $\bar{R}(k) = d_{2,k}\hat{\sigma}$.				

Example

The following simulation is $x \sim N(100, 1^2)$. Sample sizes are 3, 4, and 5.

The first sample occupies cells B4 through F4; n is COUNT(B4:F4); and $\bar{x}$ is AVERAGE(B4:F4). The spreadsheet does not treat blanks as zeros; it recognizes them as blanks. Therefore, AVERAGE(B5:F5) for the second sample returns 100.0. This means we can write the function in the top cell, and copy it down the column without worrying about the sample sizes. For the first sample, s is STDEV(B4:F4), and R is MAX(B4:F4) – MIN(B4:F4).

The variables c_4 and d_2 depend on n. Excel's LOOKUP function provides a convenient way to handle this. We put the following table in the spreadsheet. For the first sample, s/c_4 equals I4/LOOKUP (G4,A\$61:A\$64,B\$61:B\$64), where column I contains s and G contains n. The function tells Excel to find the row in A\$61:A\$64 that contains $n = 5$, and return the corresponding value in B\$61:B\$64. The \$ in the cell address tells the spreadsheet *not* to increment the row numbers when the function is copied. Similarly, R/d_2 equals K4/LOOKUP(G4, A\$61:A\$64,C\$61:C\$64).

	A	B	C
60	Sample size	c_4	d_2
61	2	0.7979	1.128
62	3	0.8862	1.693
63	4	0.9213	2.059
64	5	0.9400	2.326

(*Note:* Since the table is from a spreadsheet, the numbers are rounded off. Thus, hand calculations may not produce identical results.)

A sample calculation for the first sample is as follows: $s = 0.5755$, $R = 1.235$. For $n = 5$, $c_4 = 0.9400$, $s/c_4 = 0.6122$, and $d_2 = 2.326$, $R/d_2 = 0.531$.

The estimate for s that uses sample standard deviations is 0.950. The estimate that uses ranges is 0.942, and the standard deviation of all the data together is 0.933. These estimates are very close to each other.

Next, how would we set control limits for each sample? Here is the procedure for the $\bar{x}/s$ chart: $\bar{s} = c_{4,n}\hat{\sigma} = c_{4,n} \times 0.950$, and this is the centerline for the s chart. The control limits for the $\bar{x}$ chart are $\bar{\bar{x}} \pm A_{3,n}\bar{s} = 99.9 \pm A_{3,n}\bar{s}$.

Sample	\ Measurement 1	2	3	4	5	n	$\bar{x}$	s	s/c_4	Range R	R/d_2
1	100.607	101.040	99.822	99.805	99.805	5	100.2	0.576	0.612	1.235	0.531
2	100.342	100.125	99.554	100.104		4	100.0	0.336	0.365	0.788	0.383
3	100.553	99.469	101.732			3	100.6	1.132	1.277	2.263	1.337
4	99.933	100.369	98.873			3	99.7	0.769	0.868	1.496	0.884
5	100.435	99.726	100.175			3	100.1	0.359	0.405	0.709	0.419
6	99.717	99.734	100.649	99.318	99.318	5	99.7	0.544	0.578	1.331	0.572
7	100.375	100.307	99.389	98.473	98.473	5	99.4	0.934	0.994	1.902	0.818
8	101.320	99.987	98.975			3	100.1	1.176	1.327	2.345	1.385
9	101.387	99.272	100.550	98.222		4	99.9	1.395	1.514	3.165	1.537
10	98.291	100.553	99.676			3	99.5	1.140	1.287	2.262	1.336
11	99.855	97.556	99.950			3	99.1	1.356	1.530	2.394	1.414
12	97.772	100.343	101.351			3	99.8	1.846	2.082	3.579	2.114
13	100.272	97.715	101.013	99.143	99.143	5	99.5	1.257	1.337	3.298	1.418
14	98.457	99.086	99.253	98.837	98.837	5	98.9	0.301	0.321	0.796	0.342
15	100.749	100.055	101.493	99.098		4	100.3	1.020	1.107	2.395	1.163
16	100.178	100.268	99.981	99.333	99.333	5	99.8	0.455	0.484	0.935	0.402
17	99.583	98.934	99.887			3	99.5	0.487	0.549	0.953	0.563
18	97.568	99.943	99.878	98.809		4	99.0	1.116	1.212	2.375	1.153
19	100.759	100.935	99.863			3	100.5	0.575	0.649	1.072	0.633
20	98.840	98.089	99.708	98.544		4	98.8	0.682	0.741	1.619	0.786
21	98.880	100.567	101.099			3	100.2	1.159	1.307	2.219	1.311
22	99.609	101.308	100.617			3	100.5	0.854	0.964	1.699	1.004
23	99.786	102.671	100.942	99.490	99.490	5	100.5	1.365	1.452	3.181	1.368
24	98.168	100.192	100.433			3	99.6	1.244	1.404	2.265	1.338
25	99.622	99.690	97.635	100.354		4	99.3	1.174	1.275	2.719	1.321
26	100.353	99.127	100.948	99.939		4	100.1	0.765	0.830	1.821	0.884
27	100.444	99.148	99.593			3	99.7	0.659	0.743	1.296	0.766
28	100.237	99.402	99.732	99.388	99.388	5	99.6	0.370	0.394	0.849	0.365
29	98.546	101.616	100.681	100.546		4	100.3	1.292	1.402	3.070	1.491
30	100.489	99.203	100.646	99.471	99.471	5	99.9	0.661	0.703	1.443	0.620
31	100.742	99.939	99.687	99.506	99.506	5	99.9	0.516	0.549	1.236	0.531
32	99.173	99.438	99.941	100.130	100.130	5	99.8	0.434	0.462	0.957	0.411
33	100.358	100.313	100.425	100.137		4	100.3	0.123	0.134	0.288	0.140
34	99.988	100.029	100.261	100.632	100.632	5	100.3	0.313	0.333	0.644	0.277
35	100.658	100.931	99.893			3	100.5	0.538	0.607	1.038	0.613
36	101.382	98.803	100.535	101.311	101.311	5	100.7	1.099	1.169	2.579	1.109
37	100.459	100.006	99.989			3	100.2	0.267	0.301	0.470	0.278
38	101.027	99.236	99.636	101.205		4	100.3	0.986	1.071	1.969	0.956
39	100.556	98.782	101.836			3	100.4	1.534	1.731	3.054	1.804
40	99.270	99.766	100.860	99.894		4	99.9	0.665	0.722	1.590	0.772
41	99.988	101.835	101.505	98.599		4	100.5	1.491	1.618	3.236	1.572
42	100.921	98.777	99.014	99.157		4	99.5	0.982	1.066	2.144	1.041
43	99.046	100.120	99.705	97.856	97.856	5	98.9	1.041	1.108	2.264	0.973
44	100.055	99.946	99.841	100.951	100.951	5	100.3	0.555	0.590	1.110	0.477
45	99.963	98.170	100.739	99.456	99.456	5	99.6	0.936	0.996	2.569	1.104
46	100.304	99.030	100.140	101.029	101.029	5	100.3	0.822	0.874	1.999	0.859
47	100.119	98.409	101.658			3	100.1	1.625	1.834	3.249	1.919
48	100.064	99.743	98.805			3	99.5	0.654	0.738	1.259	0.744
49	99.922	98.608	99.981	100.618	100.618	5	99.9	0.821	0.873	2.010	0.864
50	101.185	99.461	100.696			3	100.4	0.888	1.003	1.724	1.018
Grand average	99.897	s, all data			0.933	$\hat{\sigma}$ based on s values				0.950	on R	0.942

Sample	n	$\bar{x}$	$\bar{s}$	LCL, $\bar{x}$	UCL, $\bar{x}$	s	LCL, s	UCL, s
1	5	100.2	0.893	98.63	101.17	0.576	0.000	1.865
2	4	100.0	0.875	98.48	101.32	0.336	0.000	1.983
3	3	100.6	0.842	98.25	101.55	1.132	0.000	2.162
4	3	99.7	0.842	98.25	101.55	0.769	0.000	2.162
5	3	100.1	0.842	98.25	101.55	0.359	0.000	2.162

The control limits for the s chart are $B_{3,\,n}\bar{s}$, and $B_{4,\,n}\bar{s}$. B_3 is 0 for samples of 2 to 5, so the s chart's LCL is 0. The LOOKUP function is again useful for looking up the control chart factors. Here are the results (above) for the first five samples. The control limits move to accommodate the sample size, and, for the s chart, the centerline also moves.

The calculation for the first sample is as follows: $\bar{s} = \hat{\sigma}c_4 = 0.950 \times 0.9400 = 0.893$. The control limits for $\bar{x}$ are $\bar{\bar{x}} = 99.9 \pm A_3\bar{s} = 99.9 \pm 1.427 \times 0.893$. The UCL for s is $B_4\bar{s} = 2.089 \times 0.893 = 1.865$. Since $B_3 = 0$, the LCL for s is 0. Figures 6.12 and 6.13 show the s and $\bar{x}$ charts for the data.

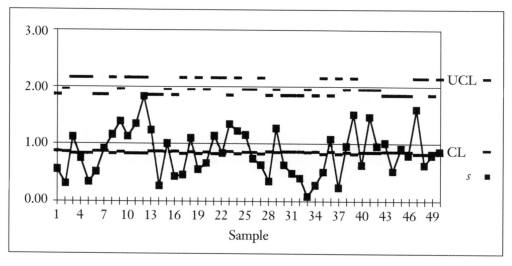

Figure 6.12. s chart, varying sample sizes.

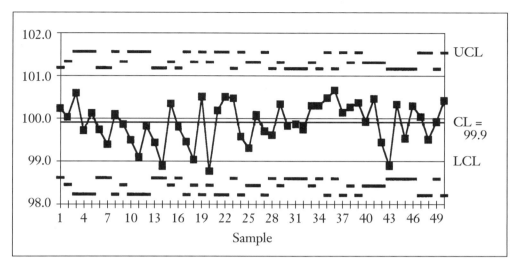

Figure 6.13. $\bar{x}$ chart, varying sample sizes.

Empirical Chart for Individuals (X Charts)

In some applications, we can get only samples of one. The X chart is the chart for process mean when the samples are individuals. To estimate the standard deviation, we can use the average moving range. For samples 1 to m,

$$MR_i = x_i - x_{i-1} \text{ for } i = 2 \text{ to } m$$

$$\overline{MR} = \frac{1}{m-1} \Sigma_{i=2}^{m} |MR_i| \text{ average moving range where } |\ | = \text{absolute value}$$
(Eq. set 6.7)

$$\hat{\sigma} = \frac{\sqrt{\pi}}{2} \overline{MR} \Rightarrow \text{control limits are } \bar{\bar{x}} \pm \frac{3\sqrt{\pi}}{2} \overline{MR} = \bar{\bar{x}} \pm 2.66 \times \overline{MR}$$

We do not make a moving range chart. Unlike a range chart, an MR chart does not provide additional information about the process variation. We use the moving ranges only to set the control limits for the X chart. Messina (1987, 132) says the X chart contains the same information that is present in the MR chart and there is correlation between the moving ranges. ASTM (1990, 97) says, "All the information in the chart for moving ranges is contained, somewhat less explicitly, in the chart for individuals." Regarding the X chart, AT&T (1985, 22) says, "Do not plot the moving ranges calculated in step (2)."

Example

For 50 data, $x \sim N(125, 1^2)$, we have the following:

Sample	X	MR
1	126.02	
2	124.66	−1.36
3	125.89	1.23
4	125.73	−0.16
5	123.52	−2.21
6	122.84	−0.68
7	126.63	3.79
8	124.95	−1.68
9	124.93	−0.02
10	124.80	−0.12
11	126.29	1.49
12	125.66	−0.63
13	124.36	−1.31
14	124.68	0.33
15	124.68	−0.01
16	126.94	2.27
17	124.29	−2.66
18	124.32	0.03
19	124.03	−0.29
20	125.34	1.31
21	123.04	−2.30
22	124.68	1.63
23	125.98	1.30
24	124.57	−1.41
25	125.11	0.54
26	125.36	0.25
27	124.00	−1.36
28	125.50	1.50
29	123.99	−1.51
30	122.11	−1.89
31	124.30	2.19
32	125.13	0.83
33	125.45	0.32
34	124.69	−0.77
35	122.87	−1.81
36	124.56	1.69
37	126.09	1.53
38	123.39	−2.69
39	126.56	3.17
40	124.77	−1.79
41	125.05	0.28
42	126.13	1.08
43	127.07	0.94
44	125.27	−1.80
45	124.88	−0.39
46	125.66	0.78
47	125.98	0.32
48	125.75	−0.22
49	124.01	−1.74
50	125.64	1.63
	$\Sigma \, |MR|$	61.24

The data are in cells B2:B51. The sum of the absolute values of the moving ranges is {SUM(ABS(C3:C51))}, where the brackets mean an array operation in Excel. Alternately, we could use the ABS function to put the absolute values in the column next to the moving ranges and then add this column. The average moving range is $\frac{61.24}{49}$ = 1.250. $\hat{\sigma}$ = $\frac{\sqrt{\pi}}{2}$1.250 = 1.108. The standard deviation of the 50 data was actually 1.080 ≈ 1.108. The grand average is 124.96 (=AVERAGE(B2:B51)), and the control limits are 124.96 ± 2.66 × 1.250 = [121.63, 128.29]. Figure 6.14 shows the resulting *X* chart.

Figure 6.14a shows the *X* chart from StatGraphics 2.0 for Windows. The control limit and centerline information have been moved under the chart. StatGraphics apparently calculates the standard deviation of the 50 data and sets the control limits at $\hat{\mu}$ ± 3$\hat{\sigma}$. The result is indistinguishable from the chart that uses the average moving range to estimate σ.

Process Capability Indices

Process capability indices measure a process' ability to meet specifications. We looked at capable and noncapable processes by subjectively comparing them to rifles and muskets. Process capability indices quantify a process' ability to meet the specification. Equation set 6.8 defines them.

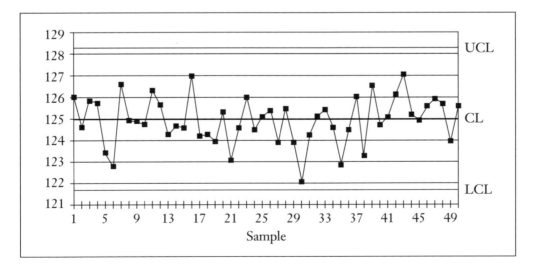

Figure 6.14. *X* chart.

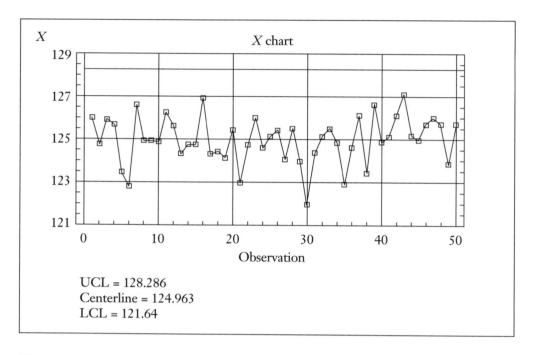

Figure 6.14a. X chart from StatGraphics 2.0.

Eq. set 6.8. Process capability indices.

Symbol	Equation	Description
C_p	$C_p = \dfrac{USL - LSL}{6\sigma}$	C_p is the ratio of the specification width to the process width. The process width is 6σ, where σ is the process standard deviation. This is also the width of the Shewhart control limits ($\pm 3\sigma$).
CPL	$CPL = \dfrac{\mu - LSL}{3\sigma}$	CPL measures the process' ability to meet the lower specification.
CPU	$CPU = \dfrac{USL - \mu}{3\sigma}$	CPU measures the process' ability to meet the upper specification.
C_{pk}	min[CPL, CPU]	C_{pk} is the minimum of CPL and CPU. When the process is at the target or nominal, CPL = CPU = C_{pk} = C_p, and the yield is at its maximum.

What do these indices mean? If the process is at its nominal, the specification limits are $3C_p$ standard deviations from the mean. We can use the cumulative normal distribution to relate C_p to yield. The nonconforming portions add to $2\Phi(-3C_p)$, so the yield is $1 - 2\Phi(-3C_p)$. If the process is not at its nominal, most of the nonconformances will be at the specification closest to the mean. This specification is $3C_{pk}$ standard deviations from the mean. The yield is $1 - \Phi(-3CPL) - \Phi(-3CPU)$. If the mean is far from the nominal, the yield is about $1 - \Phi(-3C_{pk})$. Figure 6.15 shows $2\Phi(-3C_p)$, or the nonconforming fraction for a process with its mean at the nominal. The ordinate is logarithmic. For example, a process with $C_p = 2$ will produce 1.97 nonconformances out of every billion parts. Table 6.3 shows guidelines for process capability.

Processes with $C_p < 1.33$ are not capable, and they need improvement. They are muskets instead of rifles, because their ability to hit the target is inadequate. Process improvements should focus on reducing their variation. $C_{pk} < C_p$ shows that the process is off center (not aimed at the bull's-eye). If $C_{pk} < C_p$, the process mean needs adjustment, especially if $C_{pk} \leq 1.33$.

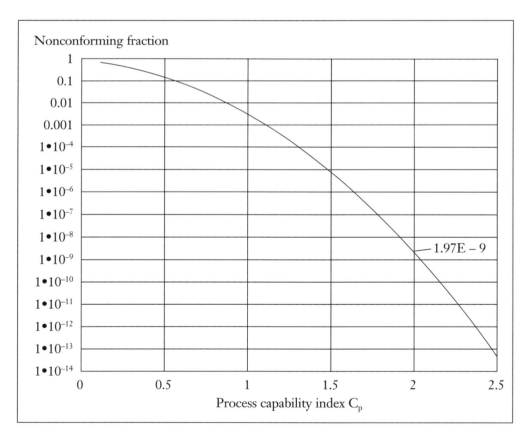

Figure 6.15. Process capability and yield.

Table 6.3. Guidelines for process capability.

C_p	Process status
$C_p < 1$	Poor. The Shewhart control limits are wider than the specification limits. The process can make bad parts even when it is in control.
$C_p < 1.33$	Fair.
$1.33 \leq C_p$	Acceptable. 1.33 is the basic standard.
$2 \leq C_p$	Excellent. The process will make less than 2 parts per billion nonconformances.

Uncertainty in Capability Indices

In practice, we do not know the process' true mean or standard deviation. We can only estimate the process capability indices.

We defined C_p as $\dfrac{\text{USL} - \text{LSL}}{6\sigma}$, but what we really have is $\hat{C}_p = \dfrac{\text{USL} - \text{LSL}}{6\hat{\sigma}}$.

The other indices include two estimated parameters: $\widehat{\text{CPL}} = \dfrac{\hat{\mu} - \text{LSL}}{3\hat{\sigma}}$ and

$\widehat{\text{CPU}} = \dfrac{\text{USL} - \hat{\mu}}{3\hat{\sigma}}$. Therefore, we don't know their exact values. We have a

point estimate such as $\hat{C}_p$, and a confidence interval around it. We are $100(1 - \alpha)$ percent sure that the true value is inside a $100(1 - \alpha)$ percent confidence interval. This interval can be very wide for process capability indices. A common mistake is to take 30 or even 50 data and then declare capability indices as absolute numbers.

The simplest confidence interval is for C_p, which relies on only one estimated parameter (σ). It uses the chi square distribution. The intervals for CPL and CPU require the noncentral t distribution, and the interval for C_{pk} is even more complex. For C_p, we are $100(1 - \alpha)$ percent sure the true value is in the following interval. Figure 6.16 is a graph of Equation 6.9.

$100(1 - \alpha)$ percent confidence interval for C_p, given n data

$$\hat{C}_p \sqrt{\frac{\chi^2_{1 - \alpha/2;\, n - 1}}{n - 1}} \leq C_p \leq \hat{C}_p \sqrt{\frac{\chi^2_{\alpha/2;\, n - 1}}{n - 1}}$$

(Eq. 6.9)

where $\chi^2_{\alpha/2;\, n - 1}$ is the $100\left(1 - \dfrac{\alpha}{2}\right)$ percentile of χ^2

with $n - 1$ degrees of freedom

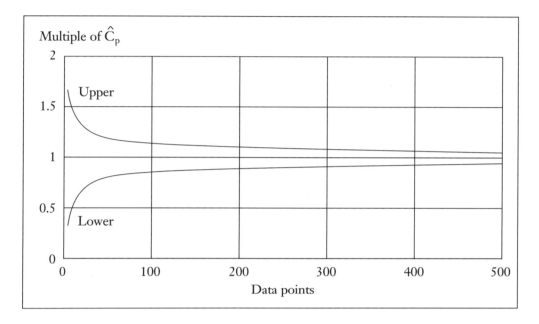

Figure 6.16. 95 percent (two-sided) confidence interval for C_p.

Even with 100 data, the 95 percent confidence interval for C_p is $[0.86\hat{C}_p, 1.14\hat{C}_p]$. We must obviously be skeptical about process capability reports whose basis is 10 or 20 data.

Appendix G shows a procedure for confidence limits for CPL and CPU. It is computation intensive and requires a computer. It is useful because it lets us put a confidence limit on the nonconforming product. The appendix also provides an approximate confidence interval for C_{pk}.

Special Control Charts

There are two control charts for special manufacturing applications. The *z chart* is for equipment that operates on different products and whose settings change frequently; and the *center band chart* or *acceptance control chart* (Feigenbaum 1991, 426–428) is for processes with an acceptable range of means. Setup limitations may prevent us from setting the tool to a specific number. For example, extremely fine adjustment may be impractical. We may have to accept a process mean within, for example, [13.02, 13.06] microns.

z Charts

The z chart is useful in job shop operations where a tool may process different products (Holmes 1988, 42). Each product may have its own specifications, nominal, and standard deviation. It is inconvenient to keep a separate chart for each product, and separate charts cannot efficiently use the Western Electric zone tests. For example, suppose product 1 generates a point in zone A. The tool next processes product 2, which also puts a point in zone A. If the points were on the same chart, they would fail the zone A test. Since they are on separate charts, we do not notice the warning of a process shift.

To plot the average of n measurements for product type j, we calculate z as follows:

$$z = \frac{\bar{x} - \mu_j}{\frac{\sigma_j}{\sqrt{n}}}$$

(Eq. 6.10)

The nominal or target for product j is μ_j, and σ_j is its standard deviation. The control limits for the z chart are simply ±3, since z is the standard normal deviate.

How can we track variation for several products on a common chart? The chi square test statistic is useful for testing the hypothesis $\sigma^2 = \sigma_0^2$, where σ_0^2 is the population's hypothetical or nominal variance. For a sample of n measurements for product j,

$$\chi^2 = \frac{(n-1)s^2}{\sigma_{0,j}^2}$$

(Eq. 6.11)

The UCL is the $100(1 - \alpha/2)$ percentile for χ^2 with $n - 1$ degrees of freedom. The acceptable risk of wrongly concluding that the process variation has increased is $\alpha/2$. The lower limit is the $100(\alpha/2)$ percentile. There is an $\alpha/2$ risk of wrongly concluding that the process variation has decreased. For Shewhart-equivalent limits, use the 0.99835 and 0.00135 quantiles respectively. (The pth quantile is the $100p$th percentile.)

Example
A tool makes three products, with nominal sizes of 100, 200, and 300 mils. The standard deviations are 1, 2, and 3 mils respectively. Here are 50 samples: 0.106 is the 0.00135 quantile of a chi square distribution with 4 degrees of freedom; 17.80 is the upper 0.99865 quantile; and 3.36 is the median, or 50th percentile. We expect half the chi squares to be on each side of it. The control charts appear in Figures 6.17 and 6.18.

i	μ_0	σ_0	\multicolumn{5}{c}{Measurement}	x bar	z	s	χ^2	\multicolumn{3}{c}{χ^2 chart}
			1	2	3	4	5					LCL	CL	UCL
1	300	3	300.21	301.25	299.91	300.15	298.81	300.07	0.05	0.869	0.34	0.106	3.36	17.80
2	300	3	299.65	299.76	297.77	296.99	293.41	297.52	−1.85	2.589	2.98	0.106	3.36	17.80
3	300	3	297.12	306.02	297.97	297.67	301.05	299.97	−0.03	3.716	6.14	0.106	3.36	17.80
4	100	1	99.14	100.83	99.40	101.13	101.04	100.31	0.69	0.958	3.67	0.106	3.36	17.80
5	100	1	101.41	102.48	101.23	101.13	98.53	100.96	2.14	1.458	8.50	0.106	3.36	17.80
6	300	3	296.79	303.49	300.58	301.47	300.71	300.61	0.45	2.432	2.63	0.106	3.36	17.80
7	300	3	303.62	302.47	301.54	300.54	298.24	301.28	0.96	2.046	1.86	0.106	3.36	17.80
8	100	1	98.59	99.60	100.72	100.78	100.84	100.11	0.24	0.987	3.90	0.106	3.36	17.80
9	100	1	99.05	100.82	98.02	100.33	99.35	99.51	−1.09	1.102	4.86	0.106	3.36	17.80
10	200	2	203.17	201.38	198.76	202.30	199.63	201.05	1.17	1.832	3.36	0.106	3.36	17.80
11	300	3	299.36	298.16	304.18	299.24	299.61	300.11	0.08	2.339	2.43	0.106	3.36	17.80
12	200	2	197.08	198.85	198.65	200.81	198.72	198.82	−1.32	1.325	1.75	0.106	3.36	17.80
13	200	2	202.12	199.72	201.81	203.20	200.70	201.51	1.69	1.341	1.80	0.106	3.36	17.80
14	300	3	301.80	294.47	297.25	299.10	301.06	298.73	−0.94	2.971	3.92	0.106	3.36	17.80
15	300	3	298.66	301.05	300.35	299.43	301.13	300.12	0.09	1.066	0.50	0.106	3.36	17.80
16	100	1	99.35	100.18	100.09	99.63	99.46	99.74	−0.58	0.372	0.55	0.106	3.36	17.80
17	100	1	100.89	99.75	100.65	100.21	99.13	100.13	0.28	0.709	2.01	0.106	3.36	17.80
18	300	3	306.32	300.66	301.56	306.20	300.60	303.07	2.29	2.937	3.83	0.106	3.36	17.80
19	300	3	297.80	301.01	293.51	297.41	302.92	298.53	−1.10	3.618	5.82	0.106	3.36	17.80
20	200	2	198.89	200.36	197.75	197.69	202.81	199.50	−0.56	2.145	4.60	0.106	3.36	17.80
21	300	3	300.66	304.98	298.17	299.43	300.70	300.79	0.59	2.561	2.91	0.106	3.36	17.80
22	100	1	101.14	100.71	100.08	98.96	99.47	100.07	0.16	0.888	3.16	0.106	3.36	17.80
23	200	2	201.37	199.27	201.64	199.69	199.59	200.31	0.35	1.105	1.22	0.106	3.36	17.80
24	300	3	305.16	299.23	294.52	299.14	298.72	299.35	−0.48	3.790	6.39	0.106	3.36	17.80
25	100	1	99.23	99.86	100.34	98.31	100.73	99.69	−0.69	0.954	3.64	0.106	3.36	17.80
26	200	2	203.92	201.20	200.79	202.22	200.09	201.64	1.84	1.487	2.21	0.106	3.36	17.80
27	100	1	100.63	99.98	100.77	100.46	100.35	100.44	0.98	0.302	0.37	0.106	3.36	17.80
28	100	1	99.85	99.55	98.40	99.32	99.38	99.30	−1.56	0.545	1.19	0.106	3.36	17.80
29	200	2	199.42	203.35	202.05	200.65	199.04	200.90	1.01	1.808	3.27	0.106	3.36	17.80
30	300	3	305.46	305.17	306.72	300.62	296.08	302.81	2.09	4.416	8.67	0.106	3.36	17.80
31	100	1	98.34	100.11	99.77	98.87	100.49	99.52	−1.08	0.888	3.16	0.106	3.36	17.80
32	300	3	299.02	299.18	297.27	301.23	298.38	299.02	−0.73	1.447	0.93	0.106	3.36	17.80
33	200	2	198.15	201.12	205.89	199.13	202.50	201.36	1.52	3.049	9.29	0.106	3.36	17.80
34	300	3	297.97	302.88	300.34	303.05	300.20	300.89	0.66	2.118	1.99	0.106	3.36	17.80
35	200	2	199.57	197.92	202.53	197.11	201.93	199.81	−0.21	2.389	5.71	0.106	3.36	17.80
36	300	3	304.48	297.42	298.23	301.32	299.87	300.26	0.20	2.794	3.47	0.106	3.36	17.80
37	100	1	100.84	100.07	99.36	98.25	100.10	99.72	−0.62	0.977	3.82	0.106	3.36	17.80
38	200	2	198.90	199.10	201.57	200.28	198.32	199.63	−0.41	1.295	1.68	0.106	3.36	17.80
39	100	1	99.82	100.59	100.91	99.44	100.13	100.18	0.40	0.589	1.39	0.106	3.36	17.80
40	200	2	199.91	199.78	201.79	202.27	201.36	201.02	1.14	1.125	1.27	0.106	3.36	17.80
41	200	2	201.62	198.43	199.23	200.10	201.17	200.11	0.12	1.325	1.75	0.106	3.36	17.80
42	100	1	101.19	99.55	101.70	100.37	99.88	100.54	1.21	0.897	3.22	0.106	3.36	17.80
43	100	1	99.38	100.62	98.64	99.62	101.25	99.90	−0.22	1.032	4.26	0.106	3.36	17.80
44	300	3	299.82	299.51	296.74	298.71	299.56	298.87	−0.85	1.259	0.70	0.106	3.36	17.80
45	100	1	100.79	98.33	98.94	99.93	97.92	99.18	−1.83	1.175	5.52	0.106	3.36	17.80
46	200	2	203.65	201.55	200.00	199.08	202.96	201.45	1.62	1.925	3.70	0.106	3.36	17.80
47	200	2	200.56	200.95	200.76	199.91	201.26	200.69	0.77	0.507	0.26	0.106	3.36	17.80
48	300	3	301.26	300.00	300.44	298.41	299.75	299.97	−0.02	1.046	0.49	0.106	3.36	17.80
49	100	1	101.36	98.96	99.86	97.76	99.91	99.57	−0.96	1.329	7.07	0.106	3.36	17.80
50	200	2	201.55	199.33	202.73	197.51	199.90	200.20	0.23	2.020	4.08	0.106	3.36	17.80

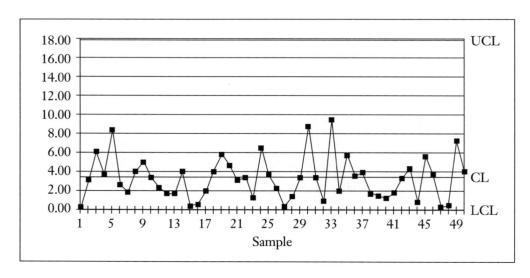

Figure 6.17. χ^2 chart for multiple products.

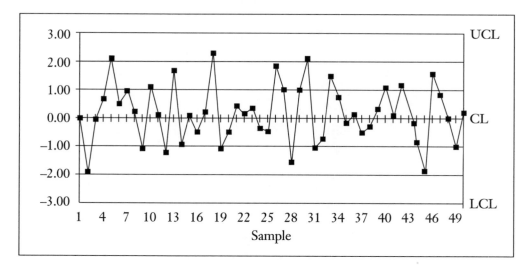

Figure 6.18. z chart for multiple products.

Center Band Charts (Acceptance Control Charts)

The center band chart is for processes that have acceptable ranges for their means. This situation may arise when it is impractical to set the process to a single mean. Suppose that a process' specification limits are [12, 14] microns. We want its mean to be 13.00 microns. In practice, we find that trying to adjust it from 13.02 to 13.00 often reduces it to 12.95 or even 12.90. Trying to change it from 12.95 to 13.00 often increases it to 13.05 or 13.10. We decide to leave it alone if the mean is between 12.9 and 13.1 microns.

Let the acceptable process mean fall within the *lower and upper acceptable process levels* [LAPL, UAPL]. A simple Shewhart chart would then look like Figure 6.19. To count toward the zone C test, a point must be outside the central band. If the mean is slightly above the centerline, there may be a run of eight points in zone C. We want to adjust the process, however, only if the mean is outside [LAPL, UAPL].

Feigenbaum (1991, 426–428) shows a procedure for customizing control charts. The underlying idea is similar to that for acceptance sampling.

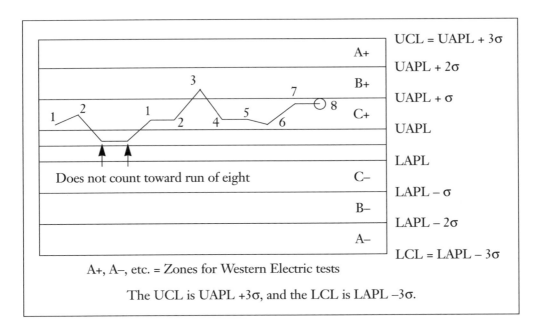

Figure 6.19. Shewhart chart with center band.

Acceptance control chart	Acceptance sampling	Risk
Upper rejectable process level URPL = USL $- z_{p2}\sigma$	Lot tolerance percent defective (LTPD)	Risk of accepting the process when it is at the URPL = 100β%. (Risk of accepting the lot when it the nonconforming fraction equals the LTPD.)
Lower rejectable process level LRPL = LSL $+ z_{p2}\sigma$	LTPD	Risk of accepting the process (lot) when it is below the LRPL = 100β%.
Upper acceptable process level UAPL = USL $- z_{p1}\sigma$	Acceptable quality level (AQL)	Risk of rejecting the process when it is below the UAPL. (Risk of rejecting the lot when the nonconforming fraction equals the AQL.)
Lower acceptable process level LAPL = LSL $+ z_{p1}\sigma$	AQL	Risk of rejecting the process when it is above the LAPL.

z_{p1} and z_{p2} are the distances, in standard deviations, between the specification limit and the corresponding APL and RPL respectively.

Then,

$$UCL = UAPL + \frac{z_{\alpha/2}\sigma}{\sqrt{n}} \text{ and } LCL = LAPL - \frac{z_{\alpha/2}\sigma}{\sqrt{n}} \text{ or}$$

$$UCL = URPL - \frac{z_{\beta/2}\sigma}{\sqrt{n}} \text{ and } LCL = LRPL + \frac{z_{\beta/2}\sigma}{\sqrt{n}}, \text{ where}$$

(Eq. set 6.12)

$$\text{sample size } n = \left[\frac{(z_\alpha + z_\beta)\sigma}{RPL - APL}\right]^2 \text{ (round up)}$$

Note that the risks apply to each specification limit.

Example

A manufacturing process' specification limits are [12, 14] microns, and the standard deviation is 0.2 microns. When the process is making 10 ppm non-conformances, we want to accept it 99 percent of the time; that is, α = 0.01. When it is making 1000 ppm nonconformances, we want a 5 percent risk of accepting it (β = 0.05). What are the control limits and sample size?

There is a false alarm risk and a risk of not rejecting an unacceptable process for each specification. We want a 0.5 percent false alarm risk for each specification and a 2.5 percent risk of accepting a bad process at each specification.

z_{p1} Standard normal deviate for $1 - 10 \times 10^{-6} = 0.99999 = 4.259$

z_{p2} Standard normal deviate for $1 - 1000 \times 10^{-6} = 0.999 = 3.090$

$z_{\alpha/2}$ Standard normal deviate for $0.995 = 2.576$

$z_{\beta/2}$ Standard normal deviate for $0.975 = 1.960$

UAPL = 12 microns − 4.259 × 0.2 microns = 11.148 microns

URPL = 12 microns − 3.090 × 0.2 microns = 11.382 microns

LAPL = 10 microns + 4.259 × 0.2 microns = 10.852 microns

LRPL = 10 microns + 3.090 × 0.2 microns = 10.618 microns

$$n = \left[\frac{(1.960 + 2.576)(0.2 \text{ microns})}{11.382 - 11.148} \right]^2 = 15.03$$

$$\text{UCL} = 11.148 \text{ microns} + \frac{2.576 \times 0.2 \text{ microns}}{\sqrt{15.03}} = 11.281 \text{ or}$$

$$\text{UCL} = 11.382 \text{ microns} - \frac{1.960 \times 0.2 \text{ microns}}{\sqrt{15.03}} = 11.281 \text{ microns}$$

The calculation is similar for the LCL. In practice, we must use a sample of 16. We round 15.03 up (not off) to make sure the sampling plan meets or exceeds the requirements.

If the process is at the UAPL, what is the chance of going over the UCL?

$$z = \frac{(11.148 - 11.281) \text{ microns}}{\frac{0.2 \text{ microns}}{\sqrt{16}}} = -2.66, \; \Phi(-2.66) = 0.0039 < 0.005$$

If the process is at the URPL, what is the chance of going over the UCL?

$$z = \frac{(11.382 - 11.281) \text{ microns}}{\frac{0.2 \text{ microns}}{\sqrt{16}}} = 2.02, \; \Phi(2.02) = 0.978 > 0.975$$

The risks do not match the specifications exactly because we rounded up the sample size. Rounding up assures that the plan will meet or exceed the requirements.

CUSUM and EWMA (Awareness)

The cumulative sum (CUSUM) chart is a form of sequential sampling. It is more powerful than the Shewhart chart for small shifts in the mean, but less powerful for large shifts. The exponentially weighted moving average (EWMA) chart is also sensitive to small shifts. Montgomery (1991) describes how to set up both charts. He also discusses changes that make CUSUM more sensitive to large shifts.

CUSUM and EWMA are philosophically similar to integral process control. The chemical process industry uses controllers that respond to shifts in process conditions. For example, if a kettle's temperature changes from its set point, the controller will adjust the heat accordingly. A proportional controller's output equals $K_c e$, where K_c is the controller gain and e is the difference (error) between the set point θ_c and the measurement θ. A proportional controller alone cannot restore θ to θ_c. There will always be a steady-state error $e = \theta_c - \theta$, or *offset* (Harriott 1964, 8). Similarly, a traditional Shewhart chart is unlikely to detect a small process shift (Montgomery 1991, 280–281).

Integral control removes the offset by responding to the integral of the error. An integral controller's output equals $\dfrac{1}{T_R} \displaystyle\int_0^t e\, dt$, where T_R is the reset time (Harriott 1964, 12). The measurement θ is continuous (like a temperature or pressure), so the controller integrates it.

In noncontinuous processes where we use SPC, we do not have a continuous measurement to integrate. Instead, we add individual measurements or sample averages. CUSUM plots the quantity $S_i = \Sigma_{j=1}^{i}(\bar{x}_j - \mu_0)$ for the ith sample, where μ_0 (like θ_c) is the target or set point. EWMA uses the weighted average of successive measurements, where recent measurements have more weight than old ones. CUSUM and EWMA, then, are philosophically similar to integral control because they add or average successive measurements.

False Alarm Risks and Power of Shewhart Control Charts

Equation set 6.13 allows us to specify false alarm risks and powers for control charts. A chart's *power* is its ability to detect an undesirable shift. It depends on the control limits, sample size, and the size of the shift. Here is an equation for the power of a standard Shewhart control chart. We let the UCL be $\mu + \dfrac{3\sigma}{\sqrt{n}}$, and we assume a process shift of $\delta\sigma$. The same reasoning applies to the LCL.

$$\Pr(\overline{x} \leq \text{UCL}) = \Phi\left(\frac{\left(\mu + \dfrac{3\sigma}{\sqrt{n}}\right) - (\mu + \delta\sigma)}{\dfrac{\sigma}{\sqrt{n}}}\right) = \Phi(3 - \delta\sqrt{n}) \quad \textbf{(Eq. 6.13)}$$

$$\text{Power } \gamma(\delta, n) = 1 - \Phi(3 - \delta\sqrt{n}) = \Pr(\overline{x} \geq \text{UCL})$$

For example, the chance of detecting a one-sigma process shift with a sample of one is $1 - \Phi(3 - 1) = 1 - 0.97725 = 0.02275$ (2.275 percent). With a sample of four, the chance is $1 - \Phi(1) = 1 - 0.8413 = 0.1587$ (15.87 percent), and for a sample of nine, it is $1 - \Phi(0) = 1 - 0.5 = 0.5$ (50 percent). Figure 6.20 shows a graph of power versus process shift.

At $\delta = 0$, $\gamma = 0.00135$ (0.135 percent). This is the risk of going over a control limit when the process is in control. Since there are two control limits, the false alarm risk is 0.27 percent. The false alarm risk is always

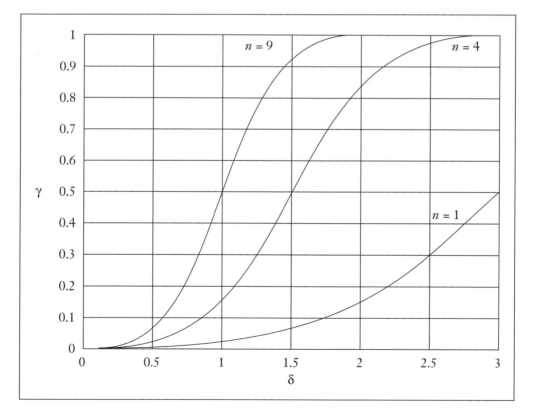

Figure 6.20. Control chart power versus process shift.

$1 - 2\Phi(3) = 0.27$ percent, no matter what the sample size is. The power to detect process shifts rises, however, with sample size. Therefore, a large sample delivers better power without increasing the false alarm risk.

The reciprocal of the power is the *average run length* (ARL). It is the number of samples we expect to measure before detecting a process shift. A plot of ARL versus the process shift appears in Figure 6.21.

We can also compute the false alarm risks and powers of the Western Electric zone tests. This calculation uses the chance that the sample average exceeds a zone limit and the cumulative binomial distribution (Equation 6.14). Table 6.4 shows the false alarm risks for the zone tests. We note the following:

- The chance that the sample average exceeds the zone limit is p (Equation set 6.15).

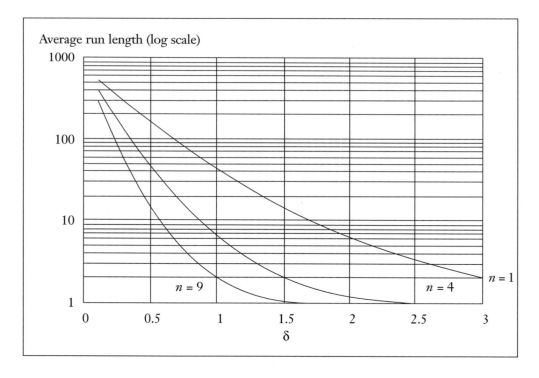

Figure 6.21. Average run length.

- The zone test looks at m samples; three for A, five for B, and eight for C.

- The samples pass the test if c or fewer exceed the zone limit, where c is the acceptance number and one for A, three for B, and 7 for C.

$$\Pr(x \leq c \mid m, p) = \Sigma_{x=0}^{c} \frac{m!}{x!(m-x)!} p^x (1-p)^{m-x} \qquad \textbf{(Eq. 6.14)}$$

Eq. set 6.15. The chance that the sample averages exceeds the zone limit.

Zone	Limit (upper section*)	p
A	$\mu + 2\dfrac{\sigma}{\sqrt{n}}$	$1 - \Phi(2 - \delta\sqrt{n})$
B	$\mu + \dfrac{\sigma}{\sqrt{n}}$	$1 - \Phi(1 - \delta\sqrt{n})$
C	μ	$1 - \Phi(-\delta\sqrt{n}) = \Phi(\delta\sqrt{n})$

*Same reasoning for the lower section

Table 6.4. False alarm risks for zone tests.

Zone	p	One-sided	Two-sided
3σ*	$1 - \Phi(3) = 0.00135$	0.00135	0.0027 (0.27%)
A	$1 - \Phi(2) = 0.02275$	$1 - \Sigma_{x=0}^{1} \dfrac{3!}{x!(3-x)!} p^x (1-p)^{3-x} = 0.00153$	0.00306 (0.31%)
B	$1 - \Phi(1) = 0.1587$	$1 - \Sigma_{x=0}^{3} \dfrac{5!}{x!(5-x)!} p^x (1-p)^{5-x} = 0.00277$	0.00544 (0.54%)
C	0.50	$1 - \Sigma_{x=0}^{7} \dfrac{8!}{x!(8-x)!} p^x (1-p)^{8-x} = 0.5^8$ $= 0.00391$	0.00782 (0.78%)
Total			1.90%

*Control limits

Therefore, using all the zone tests and the three–sigma control limits results in a 1.90 percent false alarm risk for each sample. We can also use Equation set 6.15 to calculate powers of the zone tests when the process mean has shifted. Figure 6.22 graphs the power of the zone A test.

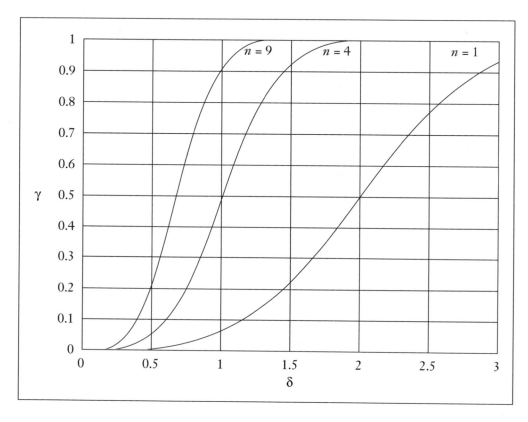

Figure 6.22. Power of the zone A test.

Percentiles of the Standard Normal Distribution*

$$\Phi(z) = \int_{-\infty}^{z} \frac{1}{\sqrt{2\pi}} \exp\left(-\frac{x^2}{2}\right) dx$$

The left column shows the ones and tenths places of the standard normal deviate. The remaining columns show the hundredths. For example, the standard normal deviate of 1.96 is $\Phi(1.96) = 0.975$.

---

*Using Microsoft Excel's NORMSDIST function.

	0.00	0.01	0.02	0.03	0.04	0.05	0.06	0.07	0.08	0.09
0.0	0.5000	0.5040	0.5080	0.5120	0.5160	0.5199	0.5239	0.5279	0.5319	0.5359
0.1	0.5398	0.5438	0.5478	0.5517	0.5557	0.5596	0.5636	0.5675	0.5714	0.5753
0.2	0.5793	0.5832	0.5871	0.5910	0.5948	0.5987	0.6026	0.6064	0.6103	0.6141
0.3	0.6179	0.6217	0.6255	0.6293	0.6331	0.6368	0.6406	0.6443	0.6480	0.6517
0.4	0.6554	0.6591	0.6628	0.6664	0.6700	0.6736	0.6772	0.6808	0.6844	0.6879
0.5	0.6915	0.6950	0.6985	0.7019	0.7054	0.7088	0.7123	0.7157	0.7190	0.7224
0.6	0.7257	0.7291	0.7324	0.7357	0.7389	0.7422	0.7454	0.7486	0.7517	0.7549
0.7	0.7580	0.7611	0.7642	0.7673	0.7704	0.7734	0.7764	0.7794	0.7823	0.7852
0.8	0.7881	0.7910	0.7939	0.7967	0.7995	0.8023	0.8051	0.8078	0.8106	0.8133
0.9	0.8159	0.8186	0.8212	0.8238	0.8264	0.8289	0.8315	0.8340	0.8365	0.8389
1.0	0.8413	0.8438	0.8461	0.8485	0.8508	0.8531	0.8554	0.8577	0.8599	0.8621
1.1	0.8643	0.8665	0.8686	0.8708	0.8729	0.8749	0.8770	0.8790	0.8810	0.8830
1.2	0.8849	0.8869	0.8888	0.8907	0.8925	0.8944	0.8962	0.8980	0.8997	0.90147
1.3	0.90320	0.90490	0.90658	0.90824	0.90988	0.91149	0.91308	0.91466	0.91621	0.91774
1.4	0.91924	0.92073	0.92220	0.92364	0.92507	0.92647	0.92785	0.92922	0.93056	0.93189
1.5	0.93319	0.93448	0.93574	0.93699	0.93822	0.93943	0.94062	0.94179	0.94295	0.94408
1.6	0.94520	0.94630	0.94738	0.94845	0.94950	0.95053	0.95154	0.95254	0.95352	0.95449
1.7	0.95543	0.95637	0.95728	0.95818	0.95907	0.95994	0.96080	0.96164	0.96246	0.96327
1.8	0.96407	0.96485	0.96562	0.96638	0.96712	0.96784	0.96856	0.96926	0.96995	0.97062
1.9	0.97128	0.97193	0.97257	0.97320	0.97381	0.97441	0.97500	0.97558	0.97615	0.97670
2.0	0.97725	0.97778	0.97831	0.97882	0.97932	0.97982	0.98030	0.98077	0.98124	0.98169
2.1	0.98214	0.98257	0.98300	0.98341	0.98382	0.98422	0.98461	0.98500	0.98537	0.98574
2.2	0.98610	0.98645	0.98679	0.98713	0.98745	0.98778	0.98809	0.98840	0.98870	0.98899
2.3	0.98928	0.98956	0.98983	0.990097	0.990358	0.990613	0.990863	0.991106	0.991344	0.991576
2.4	0.991802	0.992024	0.992240	0.992451	0.992656	0.992857	0.993053	0.993244	0.993431	0.993613
2.5	0.993790	0.993963	0.994132	0.994297	0.994457	0.994614	0.994766	0.994915	0.995060	0.995201
2.6	0.995339	0.995473	0.995603	0.995731	0.995855	0.995975	0.996093	0.996207	0.996319	0.996427
2.7	0.996533	0.996636	0.996736	0.996833	0.996928	0.997020	0.997110	0.997197	0.997282	0.997365
2.8	0.997445	0.997523	0.997599	0.997673	0.997744	0.997814	0.997882	0.997948	0.998012	0.998074
2.9	0.998134	0.998193	0.998250	0.998305	0.998359	0.998411	0.998462	0.998511	0.998559	0.998605
3.0	0.998650	0.998694	0.998736	0.998777	0.998817	0.998856	0.998893	0.998930	0.998965	0.998999
3.1	0.9990323	0.9990645	0.9990957	0.9991259	0.9991552	0.9991836	0.9992111	0.9992377	0.9992636	0.9992886
3.2	0.9993128	0.9993363	0.9993590	0.9993810	0.9994023	0.9994229	0.9994429	0.9994622	0.9994809	0.9994990
3.3	0.9995165	0.9995335	0.9995499	0.9995657	0.9995811	0.9995959	0.9996102	0.9996241	0.9996375	0.9996505
3.4	0.9996630	0.9996751	0.9996868	0.9996982	0.9997091	0.9997197	0.9997299	0.9997397	0.9997492	0.9997584
3.5	0.9997673	0.9997759	0.9997842	0.9997922	0.9997999	0.9998073	0.9998145	0.9998215	0.9998282	0.9998346
3.6	0.9998409	0.9998469	0.9998527	0.9998583	0.9998636	0.9998688	0.9998739	0.9998787	0.9998834	0.9998878
3.7	0.9998922	0.9998963	0.99990036	0.99990423	0.99990796	0.99991156	0.99991502	0.99991835	0.99992156	0.99992465
3.8	0.99992763	0.99993049	0.99993325	0.99993591	0.99993846	0.99994092	0.99994329	0.99994556	0.99994775	0.99994986
3.9	0.99995188	0.99995383	0.99995571	0.99995751	0.99995924	0.99996091	0.99996251	0.99996405	0.99996553	0.99996695
4.0	0.99996831	0.99996963	0.99997089	0.99997210	0.99997326	0.99997438	0.99997545	0.99997648	0.99997747	0.99997842

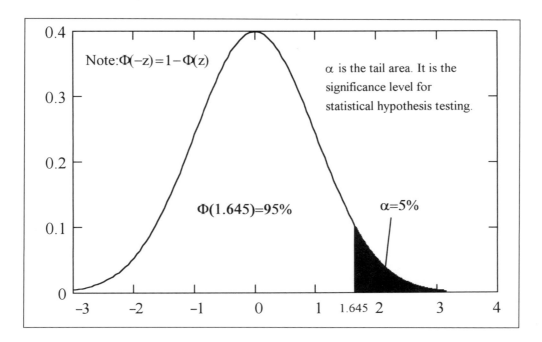

Figure A.1. Percentiles of the standard normal distribution.

APPENDIX B

Percentiles of the Chi Square Distribution*

$$F(\chi^2) = \int_0^{\chi^2} \frac{1}{2^{n/2}\Gamma\left(\frac{n}{2}\right)} \; x^{\frac{n-2}{2}} \; \exp\left(-\frac{x}{2}\right) dx \quad \text{for } n \text{ degrees of freedom}$$

---

*Using Microsoft Excel's CHIINV function.

	Cumulative chi square (*F*)												
n	0.005	0.01	0.025	0.05	0.1	0.25	0.5	0.75	0.9	0.95	0.975	0.99	0.995
1	3.93E − 05	1.57E − 04	9.82E − 04	3.93E − 03	1.58E − 02	0.102	0.455	1.32	2.71	3.84	5.02	6.63	7.88
2	1.00E − 02	2.01E − 02	5.06E − 02	0.103	0.211	10.575	1.39	2.77	4.61	5.99	7.38	9.21	10.60
3	7.17E − 02	0.115	0.216	0.352	0.584	1.21	2.37	4.11	6.25	7.81	9.35	11.34	12.84
4	0.21	0.30	0.48	0.71	1.06	1.92	3.36	5.39	7.78	9.49	11.14	13.28	14.86
5	0.41	0.55	0.83	1.15	1.61	2.67	4.35	6.63	9.24	11.07	12.83	15.09	16.75
6	0.68	0.87	1.24	1.64	2.20	3.45	5.35	7.84	10.64	12.59	14.45	16.81	18.55
7	0.99	1.24	1.69	2.17	2.83	4.25	6.35	9.04	12.02	14.07	16.01	18.48	20.28
8	1.34	1.65	2.18	2.73	3.49	5.07	7.34	10.22	13.36	15.51	17.53	20.09	21.95
9	1.73	2.09	2.70	3.33	4.17	5.90	8.34	11.39	14.68	16.92	19.02	21.67	23.59
10	2.16	2.56	3.25	3.94	4.87	6.74	9.34	12.55	15.99	18.31	20.48	23.21	25.19
11	2.60	3.05	3.82	4.57	5.58	7.58	10.34	13.70	17.28	19.68	21.92	24.73	26.76
12	3.07	3.57	4.40	5.23	6.30	8.44	11.34	14.85	18.55	21.03	23.34	26.22	28.30
13	3.57	4.11	5.01	5.89	7.04	9.30	12.34	15.98	19.81	22.36	24.74	27.69	29.82
14	4.07	4.66	5.63	6.57	7.79	10.17	13.34	17.12	21.06	23.68	26.12	29.14	31.32
15	4.60	5.23	6.26	7.26	8.55	11.04	14.34	18.25	22.31	25.00	27.49	30.58	32.80
16	5.14	5.81	6.91	7.96	9.31	11.91	15.34	19.37	23.54	26.30	28.85	32.00	34.27
17	5.70	6.41	7.56	8.67	10.09	12.79	16.34	20.49	24.77	27.59	30.19	33.41	35.72
18	6.26	7.01	8.23	9.39	10.86	13.68	17.34	21.60	25.99	28.87	31.53	34.81	37.16
19	6.84	7.63	8.91	10.12	11.65	14.56	18.34	22.72	27.20	30.14	32.85	36.19	38.58
20	7.43	8.26	9.59	10.85	12.44	15.45	19.34	23.83	28.41	31.41	34.17	37.57	40.00
21	8.03	8.90	10.28	11.59	13.24	16.34	20.34	24.93	29.62	32.67	35.48	38.93	41.40
22	8.64	9.54	10.98	12.34	14.04	17.24	21.34	26.04	30.81	33.92	36.78	40.29	42.80
23	9.26	10.20	11.69	13.09	14.85	18.14	22.34	27.14	32.01	35.17	38.08	41.64	44.18
24	9.89	10.86	12.40	13.85	15.66	19.04	23.34	28.24	33.20	36.42	39.36	42.98	45.56
25	10.52	11.52	13.12	14.61	16.47	19.94	24.34	29.34	34.38	37.65	40.65	44.31	46.93
26	11.16	12.20	13.84	15.38	17.29	20.84	25.34	30.43	35.56	38.89	41.92	45.64	48.29
27	11.81	12.88	14.57	16.15	18.11	21.75	26.34	31.53	36.74	40.11	43.19	46.96	49.65
28	12.46	13.56	15.31	16.93	18.94	22.66	27.34	32.62	37.92	41.34	44.46	48.28	50.99
29	13.12	14.26	16.05	17.71	19.77	23.57	28.34	33.71	39.09	42.56	45.72	49.59	52.34
30	13.79	14.95	16.79	18.49	20.60	24.48	29.34	34.80	40.26	43.77	46.98	50.89	53.67
31	14.46	15.66	17.54	19.28	21.43	25.39	30.34	35.89	41.42	44.99	48.23	52.19	55.00
32	15.13	16.36	18.29	20.07	22.27	26.30	31.34	36.97	42.58	46.19	49.48	53.49	56.33
33	15.82	17.07	19.05	20.87	23.11	27.22	32.34	38.06	43.75	47.40	50.73	54.78	57.65
34	16.50	17.79	19.81	21.66	23.95	28.14	33.34	39.14	44.90	48.60	51.97	56.06	58.96
35	17.19	18.51	20.57	22.47	24.80	29.05	34.34	40.22	46.06	49.80	53.20	57.34	60.27
36	17.89	19.23	21.34	23.27	25.64	29.97	35.34	41.30	47.21	51.00	54.44	58.62	61.58
37	18.59	19.96	22.11	24.07	26.49	30.89	36.34	42.38	48.36	52.19	55.67	59.89	62.88
38	19.29	20.69	22.88	24.88	27.34	31.81	37.34	43.46	49.51	53.38	56.90	61.16	64.18
39	20.00	21.43	23.65	25.70	28.20	32.74	38.34	44.54	50.66	54.57	58.12	62.43	65.48
40	20.71	22.16	24.43	26.51	29.05	33.66	39.34	45.62	51.81	55.76	59.34	63.69	66.77
41	21.42	22.91	25.21	27.33	29.91	34.58	40.34	46.69	52.95	56.94	60.56	64.95	68.05
42	22.14	23.65	26.00	28.14	30.77	35.51	41.34	47.77	54.09	58.12	61.78	66.21	69.34
43	22.86	24.40	26.79	28.96	31.63	36.44	42.34	48.84	55.23	59.30	62.99	67.46	70.62
44	23.58	25.15	27.57	29.79	32.49	37.36	43.34	49.91	56.37	60.48	64.20	68.71	71.89
45	24.31	25.90	28.37	30.61	33.35	38.29	44.34	50.98	57.51	61.66	65.41	69.96	73.17
46	25.04	26.66	29.16	31.44	34.22	39.22	45.34	52.06	58.64	62.83	66.62	71.20	74.44
47	25.77	27.42	29.96	32.27	35.08	40.15	46.34	53.13	59.77	64.00	67.82	72.44	75.70
48	26.51	28.18	30.75	33.10	35.95	41.08	47.34	54.20	60.91	65.17	69.02	73.68	76.97
49	27.25	28.94	31.55	33.93	36.82	42.01	48.33	55.27	62.04	66.34	70.22	74.92	78.23
50	27.99	29.71	32.36	34.76	37.69	42.94	49.33	56.33	63.17	67.50	71.42	76.15	79.49

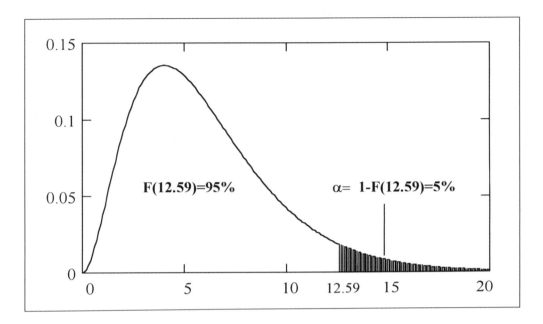

Figure B.1. Chi square distribution (6 degrees of freedom).

For $n > 30$, χ_p^2; $n \approx \frac{1}{2}\left[z_p + \sqrt{2n-1}\right]^2$ where z_p is the pth percentile of the standard normal distribution. For example, $z_{0.05} = -1.645$ and $\frac{1}{2}\left[-1.645 + \sqrt{2(50-1)}\right]^2 = 34.49$, while the fifth percentile for χ^2 with 50 degrees of freedom is 34.76. The approximation improves as n increases.

APPENDIX C

Percentiles of the Student's t Distribution*

$$F(t) = \int_{-\infty}^{z} \frac{\Gamma\left(\dfrac{n+1}{2}\right)}{\sqrt{n\pi}\ \Gamma\left(\dfrac{n}{2}\right)} \left(1 + \frac{x^2}{n}\right)^{-\frac{n+1}{2}} dx \quad \text{for } n \text{ degrees of freedom}$$

---

*Using Microsoft Excel's TINV function.

	Cumulative t distribution $F(t)$						
n	0.75	0.9	0.95	0.975	0.99	0.995	0.999
1	1.000	3.078	6.314	12.706	31.821	63.656	318.289
2	0.816	1.886	2.920	4.303	6.965	9.925	22.328
3	0.765	1.638	2.353	3.182	4.541	5.841	10.214
4	0.741	1.533	2.132	2.776	3.747	4.604	7.173
5	0.727	1.476	2.015	2.571	3.365	4.032	5.894
6	0.718	1.440	1.943	2.447	3.143	3.707	5.208
7	0.711	1.415	1.895	2.365	2.998	3.499	4.785
8	0.706	1.397	1.860	2.306	2.896	3.355	4.501
9	0.703	1.383	1.833	2.262	2.821	3.250	4.297
10	0.700	1.372	1.812	2.228	2.764	3.169	4.144
11	0.697	1.363	1.796	2.201	2.718	3.106	4.025
12	0.695	1.356	1.782	2.179	2.681	3.055	3.930
13	0.694	1.350	1.771	2.160	2.650	3.012	3.852
14	0.692	1.345	1.761	2.145	2.624	2.977	3.787
15	0.691	1.341	1.753	2.131	2.602	2.947	3.733
16	0.690	1.337	1.746	2.120	2.583	2.921	3.686
17	0.689	1.333	1.740	2.110	2.567	2.898	3.646
18	0.688	1.330	1.734	2.101	2.552	2.878	3.610
19	0.688	1.328	1.729	2.093	2.539	2.861	3.579
20	0.687	1.325	1.725	2.086	2.528	2.845	3.552
21	0.686	1.323	1.721	2.080	2.518	2.831	3.527
22	0.686	1.321	1.717	2.074	2.508	2.819	3.505
23	0.685	1.319	1.714	2.069	2.500	2.807	3.485
24	0.685	1.318	1.711	2.064	2.492	2.797	3.467
25	0.684	1.316	1.708	2.060	2.485	2.787	3.450
26	0.684	1.315	1.706	2.056	2.479	2.779	3.435
27	0.684	1.314	1.703	2.052	2.473	2.771	3.421
28	0.683	1.313	1.701	2.048	2.467	2.763	3.408
29	0.683	1.311	1.699	2.045	2.462	2.756	3.396
30	0.683	1.310	1.697	2.042	2.457	2.750	3.385
40	0.681	1.303	1.684	2.021	2.423	2.704	3.307
50	0.679	1.299	1.676	2.009	2.403	2.678	3.261
60	0.679	1.296	1.671	2.000	2.390	2.660	3.232
70	0.678	1.294	1.667	1.994	2.381	2.648	3.211
80	0.678	1.292	1.664	1.990	2.374	2.639	3.195
90	0.677	1.291	1.662	1.987	2.368	2.632	3.183
100	0.677	1.290	1.660	1.984	2.364	2.626	3.174
110	0.677	1.289	1.659	1.982	2.361	2.621	3.166
120	0.677	1.289	1.658	1.980	2.358	2.617	3.160
1000000	0.674	1.282	1.645	1.960	2.326	2.576	3.090

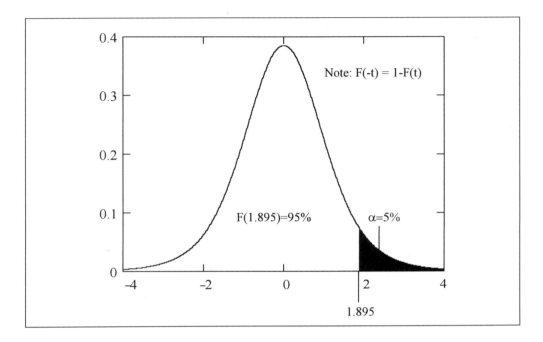

Figure C.1. *t* distribution (7 degrees of freedom).

APPENDIX D

F **Distribution Tables***

$$F(F) = \int_0^F \frac{\Gamma\left(\dfrac{m+n}{2}\right)}{\Gamma\left(\dfrac{m}{2}\right)\Gamma\left(\dfrac{n}{2}\right)} m^{\frac{m}{2}} n^{\frac{n}{2}} x^{\frac{m}{2}-1} (n + mx)^{-\frac{m+n}{2}} dx$$

for m degrees of freedom in the numerator, and n in the denominator

---

*Using Microsoft Excel's FINV function.

Table D.1. 90th percentile of the F distribution.

n	\multicolumn{18}{c}{Numerator (m)}
	1	2	3	4	5	6	7	8	9	10	12	15	20	24	30	40	60	120
1	39.86	49.50	53.59	55.83	57.24	58.20	58.91	59.44	59.86	60.19	60.71	61.22	61.74	62.00	62.26	62.53	62.79	63.06
2	8.53	9.00	9.16	9.24	9.29	9.33	9.35	9.37	9.38	9.39	9.41	9.42	9.44	9.45	9.46	9.47	9.47	9.48
3	5.54	5.46	5.39	5.34	5.31	5.28	5.27	5.25	5.24	5.23	5.22	5.20	5.18	5.18	5.17	5.16	5.15	5.14
4	4.54	4.32	4.19	4.11	4.05	4.01	3.98	3.95	3.94	3.92	3.90	3.87	3.84	3.83	3.82	3.80	3.79	3.78
5	4.06	3.78	3.62	3.52	3.45	3.40	3.37	3.34	3.32	3.30	3.27	3.24	3.21	3.19	3.17	3.16	3.14	3.12
6	3.78	3.46	3.29	3.18	3.11	3.05	3.01	2.98	2.96	2.94	2.90	2.87	2.84	2.82	2.80	2.78	2.76	2.74
7	3.59	3.26	3.07	2.96	2.88	2.83	2.78	2.75	2.72	2.70	2.67	2.63	2.59	2.58	2.56	2.54	2.51	2.49
8	3.46	3.11	2.92	2.81	2.73	2.67	2.62	2.59	2.56	2.54	2.50	2.46	2.42	2.40	2.38	2.36	2.34	2.32
9	3.36	3.01	2.81	2.69	2.61	2.55	2.51	2.47	2.44	2.42	2.38	2.34	2.30	2.28	2.25	2.23	2.21	2.18
10	3.29	2.92	2.73	2.61	2.52	2.46	2.41	2.38	2.35	2.32	2.28	2.24	2.20	2.18	2.16	2.13	2.11	2.08
11	3.23	2.86	2.66	2.54	2.45	2.39	2.34	2.30	2.27	2.25	2.21	2.17	2.12	2.10	2.08	2.05	2.03	2.00
12	3.18	2.81	2.61	2.48	2.39	2.33	2.28	2.24	2.21	2.19	2.15	2.10	2.06	2.04	2.01	1.99	1.96	1.93
13	3.14	2.76	2.56	2.43	2.35	2.28	2.23	2.20	2.16	2.14	2.10	2.05	2.01	1.98	1.96	1.93	1.90	1.88
14	3.10	2.73	2.52	2.39	2.31	2.24	2.19	2.15	2.12	2.10	2.05	2.01	1.96	1.94	1.91	1.89	1.86	1.83
15	3.07	2.70	2.49	2.36	2.27	2.21	2.16	2.12	2.09	2.06	2.02	1.97	1.92	1.90	1.87	1.85	1.82	1.79
16	3.05	2.67	2.46	2.33	2.24	2.18	2.13	2.09	2.06	2.03	1.99	1.94	1.89	1.87	1.84	1.81	1.78	1.75
17	3.03	2.64	2.44	2.31	2.22	2.15	2.10	2.06	2.03	2.00	1.96	1.91	1.86	1.84	1.81	1.78	1.75	1.72
18	3.01	2.62	2.42	2.29	2.20	2.13	2.08	2.04	2.00	1.98	1.93	1.89	1.84	1.81	1.78	1.75	1.72	1.69
19	2.99	2.61	2.40	2.27	2.18	2.11	2.06	2.02	1.98	1.96	1.91	1.86	1.81	1.79	1.76	1.73	1.70	1.67
20	2.97	2.59	2.38	2.25	2.16	2.09	2.04	2.00	1.96	1.94	1.89	1.84	1.79	1.77	1.74	1.71	1.68	1.64
21	2.96	2.57	2.36	2.23	2.14	2.08	2.02	1.98	1.95	1.92	1.87	1.83	1.78	1.75	1.72	1.69	1.66	1.62
22	2.95	2.56	2.35	2.22	2.13	2.06	2.01	1.97	1.93	1.90	1.86	1.81	1.76	1.73	1.70	1.67	1.64	1.60
23	2.94	2.55	2.34	2.21	2.11	2.05	1.99	1.95	1.92	1.89	1.84	1.80	1.74	1.72	1.69	1.66	1.62	1.59
24	2.93	2.54	2.33	2.19	2.10	2.04	1.98	1.94	1.91	1.88	1.83	1.78	1.73	1.70	1.67	1.64	1.61	1.57
25	2.92	2.53	2.32	2.18	2.09	2.02	1.97	1.93	1.89	1.87	1.82	1.77	1.72	1.69	1.66	1.63	1.59	1.56
26	2.91	2.52	2.31	2.17	2.08	2.01	1.96	1.92	1.88	1.86	1.81	1.76	1.71	1.68	1.65	1.61	1.58	1.54
27	2.90	2.51	2.30	2.17	2.07	2.00	1.95	1.91	1.87	1.85	1.80	1.75	1.70	1.67	1.64	1.60	1.57	1.53
28	2.89	2.50	2.29	2.16	2.06	2.00	1.94	1.90	1.87	1.84	1.79	1.74	1.69	1.66	1.63	1.59	1.56	1.52
29	2.89	2.50	2.28	2.15	2.06	1.99	1.93	1.89	1.86	1.83	1.78	1.73	1.68	1.65	1.62	1.58	1.55	1.51
30	2.88	2.49	2.28	2.14	2.05	1.98	1.93	1.88	1.85	1.82	1.77	1.72	1.67	1.64	1.61	1.57	1.54	1.50
40	2.84	2.44	2.23	2.09	2.00	1.93	1.87	1.83	1.79	1.76	1.71	1.66	1.61	1.57	1.54	1.51	1.47	1.42
60	2.79	2.39	2.18	2.04	1.95	1.87	1.82	1.77	1.74	1.71	1.66	1.60	1.54	1.51	1.48	1.44	1.40	1.35
120	2.75	2.35	2.13	1.99	1.90	1.82	1.77	1.72	1.68	1.65	1.60	1.55	1.48	1.45	1.41	1.37	1.32	1.26

Table D.2. 95th percentile of the *F* distribution.

n	\multicolumn{18}{c}{Numerator (*m*)}
	1	2	3	4	5	6	7	8	9	10	12	15	20	24	30	40	60	120
1	161.4	199.5	215.7	224.6	230.2	234.0	236.8	238.9	240.5	241.9	243.9	245.9	248.0	249.1	250.1	251.1	252.2	253.3
2	18.51	19.00	19.16	19.25	19.30	19.33	19.35	19.37	19.38	19.40	19.41	19.43	19.45	19.45	19.46	19.47	19.48	19.49
3	10.13	9.55	9.28	9.12	9.01	8.94	8.89	8.85	8.81	8.79	8.74	8.70	8.66	8.64	8.62	8.59	8.57	8.55
4	7.71	6.94	6.59	6.39	6.26	6.16	6.09	6.04	6.00	5.96	5.91	5.86	5.80	5.77	5.75	5.72	5.69	5.66
5	6.61	5.79	5.41	5.19	5.05	4.95	4.88	4.82	4.77	4.74	4.68	4.62	4.56	4.53	4.50	4.46	4.43	4.40
6	5.99	5.14	4.76	4.53	4.39	4.28	4.21	4.15	4.10	4.06	4.00	3.94	3.87	3.84	3.81	3.77	3.74	3.70
7	5.59	4.74	4.35	4.12	3.97	3.87	3.79	3.73	3.68	3.64	3.57	3.51	3.44	3.41	3.38	3.34	3.30	3.27
8	5.32	4.46	4.07	3.84	3.69	3.58	3.50	3.44	3.39	3.35	3.28	3.22	3.15	3.12	3.08	3.04	3.01	2.97
9	5.12	4.26	3.86	3.63	3.48	3.37	3.29	3.23	3.18	3.14	3.07	3.01	2.94	2.90	2.86	2.83	2.79	2.75
10	4.96	4.10	3.71	3.48	3.33	3.22	3.14	3.07	3.02	2.98	2.91	2.85	2.77	2.74	2.70	2.66	2.62	2.58
11	4.84	3.98	3.59	3.36	3.20	3.09	3.01	2.95	2.90	2.85	2.79	2.72	2.65	2.61	2.57	2.53	2.49	2.45
12	4.75	3.89	3.49	3.26	3.11	3.00	2.91	2.85	2.80	2.75	2.69	2.62	2.54	2.51	2.47	2.43	2.38	2.34
13	4.67	3.81	3.41	3.18	3.03	2.92	2.83	2.77	2.71	2.67	2.60	2.53	2.46	2.42	2.38	2.34	2.30	2.25
14	4.60	3.74	3.34	3.11	2.96	2.85	2.76	2.70	2.65	2.60	2.53	2.46	2.39	2.35	2.31	2.27	2.22	2.18
15	4.54	3.68	3.29	3.06	2.90	2.79	2.71	2.64	2.59	2.54	2.48	2.40	2.33	2.29	2.25	2.20	2.16	2.11
16	4.49	3.63	3.24	3.01	2.85	2.74	2.66	2.59	2.54	2.49	2.42	2.35	2.28	2.24	2.19	2.15	2.11	2.06
17	4.45	3.59	3.20	2.96	2.81	2.70	2.61	2.55	2.49	2.45	2.38	2.31	2.23	2.19	2.15	2.10	2.06	2.01
18	4.41	3.55	3.16	2.93	2.77	2.66	2.58	2.51	2.46	2.41	2.34	2.27	2.19	2.15	2.11	2.06	2.02	1.97
19	4.38	3.52	3.13	2.90	2.74	2.63	2.54	2.48	2.42	2.38	2.31	2.23	2.16	2.11	2.07	2.03	1.98	1.93
20	4.35	3.49	3.10	2.87	2.71	2.60	2.51	2.45	2.39	2.35	2.28	2.20	2.12	2.08	2.04	1.99	1.95	1.90
21	4.32	3.47	3.07	2.84	2.68	2.57	2.49	2.42	2.37	2.32	2.25	2.18	2.10	2.05	2.01	1.96	1.92	1.87
22	4.30	3.44	3.05	2.82	2.66	2.55	2.46	2.40	2.34	2.30	2.23	2.15	2.07	2.03	1.98	1.94	1.89	1.84
23	4.28	3.42	3.03	2.80	2.64	2.53	2.44	2.37	2.32	2.27	2.20	2.13	2.05	2.01	1.96	1.91	1.86	1.81
24	4.26	3.40	3.01	2.78	2.62	2.51	2.42	2.36	2.30	2.25	2.18	2.11	2.03	1.98	1.94	1.89	1.84	1.79
25	4.24	3.39	2.99	2.76	2.60	2.49	2.40	2.34	2.28	2.24	2.16	2.09	2.01	1.96	1.92	1.87	1.82	1.77
26	4.23	3.37	2.98	2.74	2.59	2.47	2.39	2.32	2.27	2.22	2.15	2.07	1.99	1.95	1.90	1.85	1.80	1.75
27	4.21	3.35	2.96	2.73	2.57	2.46	2.37	2.31	2.25	2.20	2.13	2.06	1.97	1.93	1.88	1.84	1.79	1.73
28	4.20	3.34	2.95	2.71	2.56	2.45	2.36	2.29	2.24	2.19	2.12	2.04	1.96	1.91	1.87	1.82	1.77	1.71
29	4.18	3.33	2.93	2.70	2.55	2.43	2.35	2.28	2.22	2.18	2.10	2.03	1.94	1.90	1.85	1.81	1.75	1.70
30	4.17	3.32	2.92	2.69	2.53	2.42	2.33	2.27	2.21	2.16	2.09	2.01	1.93	1.89	1.84	1.79	1.74	1.68
40	4.08	3.23	2.84	2.61	2.45	2.34	2.25	2.18	2.12	2.08	2.00	1.92	1.84	1.79	1.74	1.69	1.64	1.58
60	4.00	3.15	2.76	2.53	2.37	2.25	2.17	2.10	2.04	1.99	1.92	1.84	1.75	1.70	1.65	1.59	1.53	1.47
120	3.92	3.07	2.68	2.45	2.29	2.18	2.09	2.02	1.96	1.91	1.83	1.75	1.66	1.61	1.55	1.50	1.43	1.35

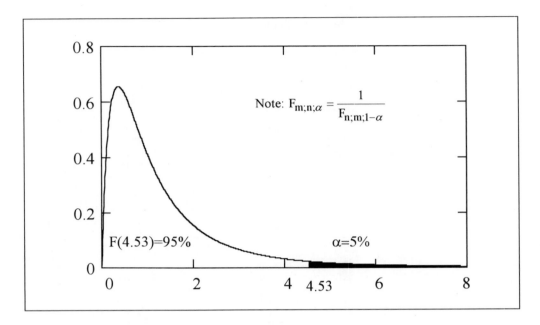

Figure D.1. *F* distribution (*m* = 4, *n* = 6 degrees of freedom).

Table D.3. 97.5th percentile of the *F* distribution.

n	\multicolumn{18}{c}{Numerator (*m*)}
	1	2	3	4	5	6	7	8	9	10	12	15	20	24	30	40	60	120
1	647.8	799.5	864.2	899.6	921.8	937.1	948.2	956.6	963.3	968.6	976.7	984.9	993.1	997.3	1001	1006	1010	1014
2	38.51	39.00	39.17	39.25	39.30	39.33	39.36	39.37	39.39	39.40	39.41	39.43	39.45	39.46	39.46	39.47	39.48	39.49
3	17.44	16.04	15.44	15.10	14.88	14.73	14.62	14.54	14.47	14.42	14.34	14.25	14.17	14.12	14.08	14.04	13.99	13.95
4	12.22	10.65	9.98	9.60	9.36	9.20	9.07	8.98	8.90	8.84	8.75	8.66	8.56	8.51	8.46	8.41	8.36	8.31
5	10.01	8.43	7.76	7.39	7.15	6.98	6.85	6.76	6.68	6.62	6.52	6.43	6.33	6.28	6.23	6.18	6.12	6.07
6	8.81	7.26	6.60	6.23	5.99	5.82	5.70	5.60	5.52	5.46	5.37	5.27	5.17	5.12	5.07	5.01	4.96	4.90
7	8.07	6.54	5.89	5.52	5.29	5.12	4.99	4.90	4.82	4.76	4.67	4.57	4.47	4.41	4.36	4.31	4.25	4.20
8	7.57	6.06	5.42	5.05	4.82	4.65	4.53	4.43	4.36	4.30	4.20	4.10	4.00	3.95	3.89	3.84	3.78	3.73
9	7.21	5.71	5.08	4.72	4.48	4.32	4.20	4.10	4.03	3.96	3.87	3.77	3.67	3.61	3.56	3.51	3.45	3.39
10	6.94	5.46	4.83	4.47	4.24	4.07	3.95	3.85	3.78	3.72	3.62	3.52	3.42	3.37	3.31	3.26	3.20	3.14
11	6.72	5.26	4.63	4.28	4.04	3.88	3.76	3.66	3.59	3.53	3.43	3.33	3.23	3.17	3.12	3.06	3.00	2.94
12	6.55	5.10	4.47	4.12	3.89	3.73	3.61	3.51	3.44	3.37	3.28	3.18	3.07	3.02	2.96	2.91	2.85	2.79
13	6.41	4.97	4.35	4.00	3.77	3.60	3.48	3.39	3.31	3.25	3.15	3.05	2.95	2.89	2.84	2.78	2.72	2.66
14	6.30	4.86	4.24	3.89	3.66	3.50	3.38	3.29	3.21	3.15	3.05	2.95	2.84	2.79	2.73	2.67	2.61	2.55
15	6.20	4.77	4.15	3.80	3.58	3.41	3.29	3.20	3.12	3.06	2.96	2.86	2.76	2.70	2.64	2.59	2.52	2.46
16	6.12	4.69	4.08	3.73	3.50	3.34	3.22	3.12	3.05	2.99	2.89	2.79	2.68	2.63	2.57	2.51	2.45	2.38
17	6.04	4.62	4.01	3.66	3.44	3.28	3.16	3.06	2.98	2.92	2.82	2.72	2.62	2.56	2.50	2.44	2.38	2.32
18	5.98	4.56	3.95	3.61	3.38	3.22	3.10	3.01	2.93	2.87	2.77	2.67	2.56	2.50	2.44	2.38	2.32	2.26
19	5.92	4.51	3.90	3.56	3.33	3.17	3.05	2.96	2.88	2.82	2.72	2.62	2.51	2.45	2.39	2.33	2.27	2.20
20	5.87	4.46	3.86	3.51	3.29	3.13	3.01	2.91	2.84	2.77	2.68	2.57	2.46	2.41	2.35	2.29	2.22	2.16
21	5.83	4.42	3.82	3.48	3.25	3.09	2.97	2.87	2.80	2.73	2.64	2.53	2.42	2.37	2.31	2.25	2.18	2.11
22	5.79	4.38	3.78	3.44	3.22	3.05	2.93	2.84	2.76	2.70	2.60	2.50	2.39	2.33	2.27	2.21	2.14	2.08
23	5.75	4.35	3.75	3.41	3.18	3.02	2.90	2.81	2.73	2.67	2.57	2.47	2.36	2.30	2.24	2.18	2.11	2.04
24	5.72	4.32	3.72	3.38	3.15	2.99	2.87	2.78	2.70	2.64	2.54	2.44	2.33	2.27	2.21	2.15	2.08	2.01
25	5.69	4.29	3.69	3.35	3.13	2.97	2.85	2.75	2.68	2.61	2.51	2.41	2.30	2.24	2.18	2.12	2.05	1.98
26	5.66	4.27	3.67	3.33	3.10	2.94	2.82	2.73	2.65	2.59	2.49	2.39	2.28	2.22	2.16	2.09	2.03	1.95
27	5.63	4.24	3.65	3.31	3.08	2.92	2.80	2.71	2.63	2.57	2.47	2.36	2.25	2.19	2.13	2.07	2.00	1.93
28	5.61	4.22	3.63	3.29	3.06	2.90	2.78	2.69	2.61	2.55	2.45	2.34	2.23	2.17	2.11	2.05	1.98	1.91
29	5.59	4.20	3.61	3.27	3.04	2.88	2.76	2.67	2.59	2.53	2.43	2.32	2.21	2.15	2.09	2.03	1.96	1.89
30	5.57	4.18	3.59	3.25	3.03	2.87	2.75	2.65	2.57	2.51	2.41	2.31	2.20	2.14	2.07	2.01	1.94	1.87
40	5.42	4.05	3.46	3.13	2.90	2.74	2.62	2.53	2.45	2.39	2.29	2.18	2.07	2.01	1.94	1.88	1.80	1.72
60	5.29	3.93	3.34	3.01	2.79	2.63	2.51	2.41	2.33	2.27	2.17	2.06	1.94	1.88	1.82	1.74	1.67	1.58
120	5.15	3.80	3.23	2.89	2.67	2.52	2.39	2.30	2.22	2.16	2.05	1.94	1.82	1.76	1.69	1.61	1.53	1.43

Table D.4. 99th percentile of the F distribution.

n	\multicolumn{18}{c}{Numerator (m)}

n	1	2	3	4	5	6	7	8	9	10	12	15	20	24	30	40	60	120
1	4052	4999	5404	5624	5764	5859	5928	5981	6022	6056	6107	6157	6209	6234	6260	6286	6313	6340
2	98.50	99.00	99.16	99.25	99.30	99.33	99.36	99.38	99.39	99.40	99.42	99.43	99.45	99.46	99.47	99.48	99.48	99.49
3	34.12	30.82	29.46	28.71	28.24	27.91	27.67	27.49	27.34	27.23	27.05	26.87	26.69	26.60	26.50	26.41	26.32	26.22
4	21.20	18.00	16.69	15.98	15.52	15.21	14.98	14.80	14.66	14.55	14.37	14.20	14.02	13.93	13.84	13.75	13.65	13.56
5	16.26	13.27	12.06	11.39	10.97	10.67	10.46	10.29	10.16	10.05	9.89	9.72	9.55	9.47	9.38	9.29	9.20	9.11
6	13.75	10.92	9.78	9.15	8.75	8.47	8.26	8.10	7.98	7.87	7.72	7.56	7.40	7.31	7.23	·7.14	7.06	6.97
7	12.25	9.55	8.45	7.85	7.46	7.19	6.99	6.84	6.72	6.62	6.47	6.31	6.16	6.07	5.99	5.91	5.82	5.74
8	11.26	8.65	7.59	7.01	6.63	6.37	6.18	6.03	5.91	5.81	5.67	5.52	5.36	5.28	5.20	5.12	5.03	4.95
9	10.56	8.02	6.99	6.42	6.06	5.80	5.61	5.47	5.35	5.26	5.11	4.96	4.81	4.73	4.65	4.57	4.48	4.40
10	10.04	7.56	6.55	5.99	5.64	5.39	5.20	5.06	4.94	4.85	4.71	4.56	4.41	4.33	4.25	4.17	4.08	4.00
11	9.65	7.21	6.22	5.67	5.32	5.07	4.89	4.74	4.63	4.54	4.40	4.25	4.10	4.02	3.94	3.86	3.78	3.69
12	9.33	6.93	5.95	5.41	5.06	4.82	4.64	4.50	4.39	4.30	4.16	4.01	3.86	3.78	3.70	3.62	3.54	3.45
13	9.07	6.70	5.74	5.21	4.86	4.62	4.44	4.30	4.19	4.10	3.96	3.82	3.66	3.59	3.51	3.43	3.34	3.25
14	8.86	6.51	5.56	5.04	4.69	4.46	4.28	4.14	4.03	3.94	3.80	3.66	3.51	3.43	3.35	3.27	3.18	3.09
15	8.68	6.36	5.42	4.89	4.56	4.32	4.14	4.00	3.89	3.80	3.67	3.52	3.37	3.29	3.21	3.13	3.05	2.96
16	8.53	6.23	5.29	4.77	4.44	4.20	4.03	3.89	3.78	3.69	3.55	3.41	3.26	3.18	3.10	3.02	2.93	2.84
17	8.40	6.11	5.19	4.67	4.34	4.10	3.93	3.79	3.68	3.59	3.46	3.31	3.16	3.08	3.00	2.92	2.83	2.75
18	8.29	6.01	5.09	4.58	4.25	4.01	3.84	3.71	3.60	3.51	3.37	3.23	3.08	3.00	2.92	2.84	2.75	2.66
19	8.18	5.93	5.01	4.50	4.17	3.94	3.77	3.63	3.52	3.43	3.30	3.15	3.00	2.92	2.84	2.76	2.67	2.58
20	8.10	5.85	4.94	4.43	4.10	3.87	3.70	3.56	3.46	3.37	3.23	3.09	2.94	2.86	2.78	2.69	2.61	2.52
21	8.02	5.78	4.87	4.37	4.04	3.81	3.64	3.51	3.40	3.31	3.17	3.03	2.88	2.80	2.72	2.64	2.55	2.46
22	7.95	5.72	4.82	4.31	3.99	3.76	3.59	3.45	3.35	3.26	3.12	2.98	2.83	2.75	2.67	2.58	2.50	2.40
23	7.88	5.66	4.76	4.26	3.94	3.71	3.54	3.41	3.30	3.21	3.07	2.93	2.78	2.70	2.62	2.54	2.45	2.35
24	7.82	5.61	4.72	4.22	3.90	3.67	3.50	3.36	3.26	3.17	3.03	2.89	2.74	2.66	2.58	2.49	2.40	2.31
25	7.77	5.57	4.68	4.18	3.85	3.63	3.46	3.32	3.22	3.13	2.99	2.85	2.70	2.62	2.54	2.45	2.36	2.27
26	7.72	5.53	4.64	4.14	3.82	3.59	3.42	3.29	3.18	3.09	2.96	2.81	2.66	2.58	2.50	2.42	2.33	2.23
27	7.68	5.49	4.60	4.11	3.78	3.56	3.39	3.26	3.15	3.06	2.93	2.78	2.63	2.55	2.47	2.38	2.29	2.20
28	7.64	5.45	4.57	4.07	3.75	3.53	3.36	3.23	3.12	3.03	2.90	2.75	2.60	2.52	2.44	2.35	2.26	2.17
29	7.60	5.42	4.54	4.04	3.73	3.50	3.33	3.20	3.09	3.00	2.87	2.73	2.57	2.49	2.41	2.33	2.23	2.14
30	7.56	5.39	4.51	4.02	3.70	3.47	3.30	3.17	3.07	2.98	2.84	2.70	2.55	2.47	2.39	2.30	2.21	2.11
40	7.31	5.18	4.31	3.83	3.51	3.29	3.12	2.99	2.89	2.80	2.66	2.52	2.37	2.29	2.20	2.11	2.02	1.92
60	7.08	4.98	4.13	3.65	3.34	3.12	2.95	2.82	2.72	2.63	2.50	2.35	2.20	2.12	2.03	1.94	1.84	1.73
120	6.85	4.79	3.95	3.48	3.17	2.96	2.79	2.66	2.56	2.47	2.34	2.19	2.03	1.95	1.86	1.76	1.66	1.53

Table D.5. 99.5th percentile of the F distribution.

n								Numerator (m)										
	1	2	3	4	5	6	7	8	9	10	12	15	20	24	30	40	60	120
1	16212	19997	21614	22501	23056	23440	23715	23924	24091	24222	24427	24632	24837	24937	25041	25146	25254	25358
2	198.5	199.0	199.2	199.2	199.3	199.3	199.4	199.4	199.4	199.4	199.4	199.4	199.4	199.4	199.5	199.5	199.5	199.5
3	55.55	49.80	47.47	46.20	45.39	44.84	44.43	44.13	43.88	43.68	43.39	43.08	42.78	42.62	42.47	42.31	42.15	41.99
4	31.33	26.28	24.26	23.15	22.46	21.98	21.62	21.35	21.14	20.97	20.70	20.44	20.17	20.03	19.89	19.75	19.61	19.47
5	22.78	18.31	16.53	15.56	14.94	14.51	14.20	13.96	13.77	13.62	13.38	13.15	12.90	12.78	12.66	12.53	12.40	12.27
6	18.63	14.54	12.92	12.03	11.46	11.07	10.79	10.57	10.39	10.25	10.03	9.81	9.59	9.47	9.36	9.24	9.12	9.00
7	16.24	12.40	10.88	10.05	9.52	9.16	8.89	8.68	8.51	8.38	8.18	7.97	7.75	7.64	7.53	7.42	7.31	7.19
8	14.69	11.04	9.60	8.81	8.30	7.95	7.69	7.50	7.34	7.21	7.01	6.81	6.61	6.50	6.40	6.29	6.18	6.06
9	13.61	10.11	8.72	7.96	7.47	7.13	6.88	6.69	6.54	6.42	6.23	6.03	5.83	5.73	5.62	5.52	5.41	5.30
10	12.83	9.43	8.08	7.34	6.87	6.54	6.30	6.12	5.97	5.85	5.66	5.47	5.27	5.17	5.07	4.97	4.86	4.75
11	12.23	8.91	7.60	6.88	6.42	6.10	5.86	5.68	5.54	5.42	5.24	5.05	4.86	4.76	4.65	4.55	4.45	4.34
12	11.75	8.51	7.23	6.52	6.07	5.76	5.52	5.35	5.20	5.09	4.91	4.72	4.53	4.43	4.33	4.23	4.12	4.01
13	11.37	8.19	6.93	6.23	5.79	5.48	5.25	5.08	4.94	4.82	4.64	4.46	4.27	4.17	4.07	3.97	3.87	3.76
14	11.06	7.92	6.68	6.00	5.56	5.26	5.03	4.86	4.72	4.60	4.43	4.25	4.06	3.96	3.86	3.76	3.66	3.55
15	10.80	7.70	6.48	5.80	5.37	5.07	4.85	4.67	4.54	4.42	4.25	4.07	3.88	3.79	3.69	3.59	3.48	3.37
16	10.58	7.51	6.30	5.64	5.21	4.91	4.69	4.52	4.38	4.27	4.10	3.92	3.73	3.64	3.54	3.44	3.33	3.22
17	10.38	7.35	6.16	5.50	5.07	4.78	4.56	4.39	4.25	4.14	3.97	3.79	3.61	3.51	3.41	3.31	3.21	3.10
18	10.22	7.21	6.03	5.37	4.96	4.66	4.44	4.28	4.14	4.03	3.86	3.68	3.50	3.40	3.30	3.20	3.10	2.99
19	10.07	7.09	5.92	5.27	4.85	4.56	4.34	4.18	4.04	3.93	3.76	3.59	3.40	3.31	3.21	3.11	3.00	2.89
20	9.94	6.99	5.82	5.17	4.76	4.47	4.26	4.09	3.96	3.85	3.68	3.50	3.32	3.22	3.12	3.02	2.92	2.81
21	9.83	6.89	5.73	5.09	4.68	4.39	4.18	4.01	3.88	3.77	3.60	3.43	3.24	3.15	3.05	2.95	2.84	2.73
22	9.73	6.81	5.65	5.02	4.61	4.32	4.11	3.94	3.81	3.70	3.54	3.36	3.18	3.08	2.98	2.88	2.77	2.66
23	9.63	6.73	5.58	4.95	4.54	4.26	4.05	3.88	3.75	3.64	3.47	3.30	3.12	3.02	2.92	2.82	2.71	2.60
24	9.55	6.66	5.52	4.89	4.49	4.20	3.99	3.83	3.69	3.59	3.42	3.25	3.06	2.97	2.87	2.77	2.66	2.55
25	9.48	6.60	5.46	4.84	4.43	4.15	3.94	3.78	3.64	3.54	3.37	3.20	3.01	2.92	2.82	2.72	2.61	2.50
26	9.41	6.54	5.41	4.79	4.38	4.10	3.89	3.73	3.60	3.49	3.33	3.15	2.97	2.87	2.77	2.67	2.56	2.45
27	9.34	6.49	5.36	4.74	4.34	4.06	3.85	3.69	3.56	3.45	3.28	3.11	2.93	2.83	2.73	2.63	2.52	2.41
28	9.28	6.44	5.32	4.70	4.30	4.02	3.81	3.65	3.52	3.41	3.25	3.07	2.89	2.79	2.69	2.59	2.48	2.37
29	9.23	6.40	5.28	4.66	4.26	3.98	3.77	3.61	3.48	3.38	3.21	3.04	2.86	2.76	2.66	2.56	2.45	2.33
30	9.18	6.35	5.24	4.62	4.23	3.95	3.74	3.58	3.45	3.34	3.18	3.01	2.82	2.73	2.63	2.52	2.42	2.30
40	8.83	6.07	4.98	4.37	3.99	3.71	3.51	3.35	3.22	3.12	2.95	2.78	2.60	2.50	2.40	2.30	2.18	2.06
60	8.49	5.79	4.73	4.14	3.76	3.49	3.29	3.13	3.01	2.90	2.74	2.57	2.39	2.29	2.19	2.08	1.96	1.83
120	8.18	5.54	4.50	3.92	3.55	3.28	3.09	2.93	2.81	2.71	2.54	2.37	2.19	2.09	1.98	1.87	1.75	1.61

Control Chart Formulas and Control Chart Factors

Control limits (LCL, UCL) for samples of n measurements

	Estimate for process variation (Given)		
	Actual standard deviation, σ	Average standard deviation, $\bar{s}$, for sample of n	Average range, $\bar{R}$, for sample of n
Chart for process mean (centerline is $\bar{\bar{x}}$ or μ).	$\mu \pm 3\dfrac{\sigma}{\sqrt{n}}$	$\bar{\bar{x}} \pm A_3\bar{s}$	$\bar{\bar{x}} \pm A_2\bar{R}$
Chart for process variation	s chart: $[B_5\sigma, B_6\sigma]$ Centerline, $c_4\sigma$ R chart: $[D_1\sigma, D_2\sigma]$ Centerline, $d_2\sigma$	s chart: $[B_3\bar{s}, B_4\bar{s}]$ Centerline, $\bar{s}$	R chart: $[D_3\bar{R}, D_4\bar{R}]$ Centerline, $\bar{R}$
Estimate for the process standard deviation σ	Given	For m samples, where the ith sample size is n_i, and $c_4(n)$ is c_4 for a sample of n, $\hat{\sigma} = \dfrac{1}{m}\,\Sigma_{i=1}^{m}\dfrac{s_i}{c_4(n_i)}$	For m samples, where the ith sample size is n_i, and $d_2(n)$ is d_2 for a sample of n, $\hat{\sigma} = \dfrac{1}{m}\,\Sigma_{i=1}^{m}\dfrac{R_i}{d_2(n_i)}$

Control Chart Factors

n	Factors for $\bar{x}$ chart			Factors for central line (s charts)		Factors for s chart				Factors for central line (range charts)			Factors for R chart			
						Given $\bar{s}$		Given σ					Given σ		Given $\bar{R}$	
	A	A_2	A_3	c_4	$1/c_4$	B_3	B_4	B_5	B_6	d_2	d_3	$1/d_2$	D_1	D_2	D_3	D_4
2	2.121	1.880	2.659	0.7979	1.2533	0.000	3.267	0.000	2.606	1.128	0.853	0.8862	0.000	3.686	0.000	3.267
3	1.732	1.023	1.954	0.8862	1.1284	0.000	2.568	0.000	2.276	1.693	0.888	0.5908	0.000	4.358	0.000	2.575
4	1.500	0.729	1.628	0.9213	1.0854	0.000	2.266	0.000	2.088	2.059	0.880	0.4857	0.000	4.698	0.000	2.282
5	1.342	0.577	1.427	0.9400	1.0638	0.000	2.089	0.000	1.964	2.326	0.864	0.4299	0.000	4.918	0.000	2.114
6	1.225	0.483	1.287	0.9515	1.0509	0.030	1.970	0.029	1.874	2.534	0.848	0.3946	0.000	5.079	0.000	2.004
7	1.134	0.419	1.182	0.9594	1.0424	0.118	1.882	0.113	1.806	2.704	0.833	0.3698	0.205	5.204	0.076	1.924
8	1.061	0.373	1.099	0.9650	1.0362	0.185	1.815	0.179	1.751	2.847	0.820	0.3512	0.388	5.307	0.136	1.864
9	1.000	0.337	1.032	0.9693	1.0317	0.239	1.761	0.232	1.707	2.970	0.808	0.3367	0.547	5.394	0.184	1.816
10	0.949	0.308	0.975	0.9727	1.0281	0.284	1.716	0.276	1.669	3.078	0.797	0.3249	0.686	5.469	0.223	1.777
11	0.905	0.285	0.927	0.9754	1.0253	0.321	1.679	0.313	1.637	3.173	0.787	0.3152	0.811	5.535	0.256	1.744
12	0.866	0.266	0.886	0.9776	1.0230	0.354	1.646	0.346	1.610	3.258	0.778	0.3069	0.923	5.594	0.283	1.717
13	0.832	0.249	0.850	0.9794	1.0210	0.382	1.618	0.374	1.585	3.336	0.770	0.2998	1.025	5.647	0.307	1.693
14	0.802	0.235	0.817	0.9810	1.0194	0.406	1.594	0.398	1.563	3.407	0.763	0.2935	1.118	5.696	0.328	1.672
15	0.775	0.223	0.789	0.9823	1.0180	0.428	1.572	0.421	1.544	3.472	0.756	0.2880	1.203	5.740	0.347	1.653
16	0.750	0.212	0.763	0.9835	1.0168	0.448	1.552	0.440	1.527	3.532	0.750	0.2831	1.282	5.782	0.363	1.637
17	0.728	0.203	0.739	0.9845	1.0157	0.466	1.534	0.459	1.510	3.588	0.744	0.2787	1.356	5.820	0.378	1.622
18	0.707	0.194	0.718	0.9854	1.0148	0.482	1.518	0.475	1.496	3.640	0.739	0.2747	1.424	5.856	0.391	1.609
19	0.688	0.187	0.698	0.9862	1.0140	0.497	1.503	0.490	1.483	3.689	0.733	0.2711	1.489	5.889	0.404	1.596
20	0.671	0.180	0.680	0.9869	1.0132	0.510	1.490	0.503	1.470	3.735	0.729	0.2677	1.549	5.921	0.415	1.585
21	0.655	0.173	0.663	0.9876	1.0126	0.523	1.477	0.516	1.459	3.778	0.724	0.2647	1.606	5.951	0.425	1.575
22	0.640	0.167	0.647	0.9882	1.0120	0.534	1.466	0.528	1.448	3.819	0.720	0.2618	1.660	5.979	0.435	1.565
23	0.626	0.162	0.633	0.9887	1.0114	0.545	1.455	0.539	1.438	3.858	0.716	0.2592	1.711	6.006	0.443	1.557
24	0.612	0.157	0.619	0.9892	1.0109	0.555	1.445	0.549	1.429	3.895	0.712	0.2567	1.759	6.032	0.452	1.548
25	0.600	0.153	0.606	0.9896	1.0105	0.565	1.435	0.559	1.420	3.931	0.708	0.2544	1.805	6.056	0.459	1.541

Formulas for Control Chart Factors*

These are needed only if it is necessary to program a computer to calculate the factors.

$c_4 = \sqrt{\dfrac{2}{n-1}}\dfrac{\left(\dfrac{n-2}{2}\right)!}{\left(\dfrac{n-3}{2}\right)!} = \sqrt{\dfrac{2}{n-1}}\dfrac{\Gamma\left(\dfrac{n}{2}\right)}{\Gamma\left(\dfrac{n-1}{2}\right)}$	$B_5 = c_4 - 3\sqrt{1 - c_4^2}$ $B_6 = c_4 + 3\sqrt{1 - c_4^2}$
$A = \dfrac{3}{\sqrt{n}}$	$B_3 = \dfrac{B_5}{c_4} \quad B_4 = \dfrac{B_6}{c_4}$
$A_2 = \dfrac{3}{d_2\sqrt{n}}$	$D_1 = d_2 - 3d_3 \quad$ See below for d_2, d_3 $D_1 = d_2 + 3d_3$
$A_3 = \dfrac{3}{c_4\sqrt{n}}$	$D_3 = \dfrac{D_1}{d_2} \quad D_4 = \dfrac{D_2}{d_2}$

The formulas for d_2 and d_3 are computation intensive, but a 486 or Pentium processor should be adequate to reproduce their results.

$d_2 = \int_{-\infty}^{\infty}\left[1 - (1 - \Phi(x))^n - \Phi(x)^n\right]dx$ where $\Phi(x)$ is the cumulative normal function.

$d_3 = \left[2\int_{-\infty}^{\infty}\int_{-\infty}^{x_1}\left[1 - \Phi(x_1)^n - (1 - \Phi(x_n))^n + (\Phi(x_1) - \Phi(x_n))^n\right]dx_n dx_1 - d_2^2\right]^{\frac{1}{2}}$

∞ = 5 was adequate for MathCAD to reproduce tabulated results (ASTM 1990, Table 49)

---

*See ASTM 1990, 91–95.

APPENDIX F

Statistical Functions for Spreadsheets and Random Number Generation for In-House Training

In most cases, Excel and Quattro Pro use the same function names.

Purpose	Function name (@FUNCTION)		
	Microsoft Excel 5.0	Corel Quattro Pro 6.0	Lotus 1-2-3 Version 4
Cumulative standard normal distribution $\Phi(z)$	NORMSDIST	NORMSDIST	NORMAL
Inverse cumulative standard normal distribution	NORMSINV	NORMSINV	NORMAL
Inverse cumulative chi square (Find the pth percentile of a chi square distribution with n degrees of freedom.)	CHIINV	CHINV	CHIDIST
Inverse cumulative t distribution	TINV	TINV	TDIST
Inverse cumulative F distribution	FINV	FINV	FDIST
Binomial distribution	BINOMDIST	BINOMDIST	BINOMIAL
Poisson distribution	POISSON	POISSON	POISSON
Inverse cumulative binomial distribution	CRITBINOM	CRITBINOM	CRIT-BINOMIAL
Gamma function, $\Gamma(x)$			GAMMA

Random Number Generation

This appendix is useful for instructors who want to create examples for in-house training. These examples can simulate the factory's manufacturing processes and might be more relevant to students than textbook examples. They may also be useful for testing statistical software.

Microsoft Excel has a random number generation tool under its TOOLS/DATA ANALYSIS menu. The user can select the normal, binomial, and Poisson distributions, among others. For example, we can produce normally distributed data for Shewhart charts, binomial data for multiple attribute charts, and so on. The tool fills a selected range of cells with random numbers from the specified distribution. Quattro Pro and Lotus should have similar functions.

If the available software can generate random numbers only from the uniform distribution, we can still simulate the normal, binomial, and Poisson distributions. Let $a \sim U(0, 1)$; that is, a is a random number from the interval [0, 1], where every value (continuous scale) has an equal chance of occurrence. In MathCAD, rnd(1) does this.

Inverse Transform

This works for all distributions.

$F(x)$, the cumulative probability distribution for any probability density function, falls in the range [0, 1]. We can generate $a \sim U(0, 1)$ and calculate x such that $F(x) = a$. Then x is a random number from the probability density function $f(x)$. In MathCAD, root($F(x)$ – rnd(1), x) finds x such that $F(x) - (a \sim U(0, 1)) = 0$. This may take some time, since MathCAD uses an iterative procedure to find the root of the equation. Also, it will not work if x is discrete (binomial, Poisson).

We can, however, generate a table of cumulative terms for the binomial or Poisson distribution. For example, here are the cumulative terms for $p = 0.15$ and $n = 5$. We generate $a \sim U(0, 1)$, and then select x such that $F(x - 1) < a \leq F(x)$.

x	$F(x)$
0	0.4437
1	0.8352
2	0.9734
3	0.9978
4	0.9999
5	1.0000

If $a \leq 0.4437$, $x = 0$; if $0.9734 < a \leq 0.9978$, $x = 3$; if $a > 0.9999$, $x = 5$, and so on.

Normal Distribution

The Box-Muller procedure (Law and Kelton 1982, 258–259) is easy to use. We generate a_1 and $a_2 \sim U(0, 1)$. Then $x = \mu + \sigma(\sqrt{-2\ln a_1})\cos(2\pi a_2)$ where $x \sim N(0, \sigma^2)$.

Binomial Distribution

For a sample of n, $x = \Sigma_{i=1}^{n}$ (1 if $a_i \leq p$, else 0). That is, we generate n $a_i s$ (Bernoulli trials), and count those $\leq p$, where p is the chance of an event. For example, if $p = 0.07$, count 1 if $a_i \leq 0.07$, and 0 if $a_i > 0.07$.

Poisson Distribution

Law and Kelton (1982, 267) provide a method whose basis is the relationship between the Poisson and exponential distributions. To generate a random x from the Poisson distribution with mean μ, let $\Pi_{i=0}^{x}$, $a_i < e^{-\mu}$. That is, iterate until the product of the uniform random numbers is less than $e^{-\mu}$, then $x = i$, noting that i starts at zero.

Example

The following MathCAD 6 Plus algorithm uses the while loop to simulate 500 random numbers from a Poisson distribution with $\mu = 1$.

$$x_i = \begin{vmatrix} j \leftarrow 0 \\ b \leftarrow \mathrm{rnd}(1) \\ \text{while } b > \exp((-1)) \\ \quad \begin{vmatrix} b \leftarrow b \cdot \mathrm{rnd}(1) \\ j \leftarrow j + 1 \end{vmatrix} \\ j \end{vmatrix}$$

Results

x	Actual	Expected	χ^2 (5 − 1 d.f.)
0	175	184	0.440
1	189	184	0.136
2	93	92	0.011
3	31	30.5	0.008
4	10	7.5	
≥ 5	2	2	0.658
	500	500	1.253

These results easily pass the chi square test for goodness of fit.

APPENDIX G

Confidence Limits for Capability Indices CPL and CPU

This procedure uses the cumulative noncentral t distribution derived from Johnson and Kotz (1970, 204, eq. (5)). But their equation is for $\Pr(0 \le t'_\upsilon \le t)$, not $\Pr(t'_\upsilon \le t)$. Also, $e^{\frac{x^2}{2}}$ should be $e^{-\frac{x^2}{2}}$ in this equation.

$$F(t') = \frac{1}{2^{\frac{\upsilon}{2}-1}\Gamma\left(\frac{\upsilon}{2}\right)} \int_0^z x^{\upsilon-1} e^{-\frac{1}{2}x^2} \Phi\left(\frac{t'x}{\sqrt{\upsilon}} - \delta\right) dx$$

where **(Eq. G.1)**

$$\Phi(y) = \int_{-\infty}^y \frac{1}{\sqrt{2\pi}} e^{-\frac{1}{2}z^2}\ dz \text{ and } \Gamma(\alpha) = \int_0^\infty y^{\alpha-1} e^{-y} dy$$

If x is an integer, $\Gamma(\alpha) = (\alpha - 1)!$

Procedure

1. The specification limit of interest is k standard deviations from the grand average, $\bar{\bar{x}}$, of n data.

2. $t' = kn^{0.5}$.

3. Solve equation G.1 for the noncentrality parameter δ that will make $F(t') = 1 - \alpha$. This requires a procedure for finding the roots of an equation, like MathCAD's root function.

4. $\delta = zn^{0.5}$, so $z = \delta \div n^{0.5}$, and we are $100(1 - \alpha)$ percent sure that the specification is z standard deviations from the true process mean. We are $100(1 - \alpha)$ percent sure that no more than $\Phi(-z)$ of the product will be outside the specification limit.

5. The $100(1 - \alpha)$ percent lower confidence limit for CPL or CPU is one-third of this number. For example, if the lower specification is 4 standard deviations from the mean, CPL = 4/3.

Example

A sample of 30 pieces yields an estimate of 1.341 for CPL. What is the lower 95 percent confidence limit for $\overset{\wedge}{CPL}$?

The specification limit is $3\overset{\wedge}{CPL}$ = 4.022 standard deviations from the average. $t' = 4.022 \times 30^{0.5} = 22.03$. Using $\infty = 25$, a guess of 1 for CPL, or $\delta = 3 \times 30^{0.5} = 16.43$, and tolerance 10^{-6},

$$F(t, \delta, \nu) = \frac{1}{2^{\left(\frac{\nu-1}{2}\right)} \cdot \Gamma\left(\frac{\nu}{2}\right)} \cdot \int_0^{\infty} x^{\nu-1} \cdot e^{-\frac{x^2}{2}} \left(\Phi\left(\frac{t \cdot x}{\sqrt{\nu}} - \delta\right)\right) dx$$

$$\text{root } (F(4.022 \cdot \sqrt{30}, \delta, 30 - 1) - 0.95, \delta) = 16.925$$

Therefore, we are 95 percent sure the LSL is at least $16.925 \div 30^{0.5} = 3.090$ standard deviations from the process mean, and that no more than $\Phi(-3.090) = 0.0010$ (0.10 percent) of the product will be below the LSL. We are also 95 percent sure that CPL is 1.030 or greater.

To test this procedure, we use the one-sided tolerance factors in Juran and Gryna (1988, Table V, AII.36). *P* is the fraction of product within *k* standard deviations of the average $\bar{x}$. See *n* = 30, *P* = 0.999, and γ (confidence) = 0.95 for this example.

By providing *P*, and thus $\delta = z_p n^{0.5}$, and solving Equation G.1 for *t'* instead of for δ, we can calculate one-sided tolerance factors *k* as a function of $\gamma = (1 - \alpha)$ and *P*.

Franklin and Wasserman (1992), citing Bissell (1990), offer the following approximation for a lower confidence limit for C_{pk}. It is computationally simple, but it requires a moderate or large amount of data (*n* measurements) to be reliable. We shouldn't base statements about C_p or C_{pk} on small quantities of data anyway. $z_{1-\alpha}$ is the standard normal deviate for the $1 - \alpha$ quantile of the standard normal distribution. $\hat{C}_{pk}$ is the point estimate for C_{pk}, and we are $100(1 - \alpha)$ percent sure that $C_{pk} \geq c_k$.

$$c_k \approx \hat{C}_{pk} - z_{1-\alpha} \sqrt{\frac{1}{9n} + \frac{\hat{C}_{pk}^2}{2n - 2}}$$

This estimate applies when $\overset{\wedge}{CPL} \approx \overset{\wedge}{CPU}$; that is, the process mean is midway between the specifications. When the sample *n* is fairly large, and $\overset{\wedge}{CPL}$ and $\overset{\wedge}{CPU}$ differ significantly, the lower confidence limit for the smaller of CPL and CPU is an adequate estimate of c_k (Kushler and Hurley 1992).

APPENDIX H

Useful World Wide Web Resources*

Resource	Uniform resource locator (URL)
American National Standards Institute (ANSI)	http://www.ansi.org/
American Society for Quality Control	http://www.ASQC.org/
American Society for Testing and Materials (ASTM)	http://www.astm.org/stand.html
CARL (Electronic literature search vendor. No charge to search, but there is a charge for fax delivery of the articles.)	http://www.carl.org/uncover/unchome.html
Covey Leadership Center	http://www2.covey.com/covey/covey.html
Harris Semiconductor (Mountaintop)	http://www.mtp.semi.harris.com/
John Wiley & Sons (publisher of many statistics and engineering books	http://www.wiley.com/ProductInfo.html
Juran Institute	http://www.juran.com/juran/
McGraw-Hill (publisher of many engineering books)	http://www.infor.com:53311/mghp/mgh2browse.shtml
National Institute of Standards and Technology (NIST): Headquarters	http://ts.nist.gov/ts/htdocs/200.html
National Institute of Standards and Technology: Quality Programs (Baldrige)	http://www.nist.gov/quality_program/
National Institute of Standards and Technology: Reference Data Products	http://www.nist.gov/srd/srd.html
Occupational Safety and Health Administration (OSHA)	http://198.175.40/OCIS/standards_related.html
Quality Magazine	http://qualitymag.com/
Society of Manufacturing Engineers	http://www.sme.org/

*As of May 1996. Web addresses are subject to change.

Bibliography

ANSI/ISO/ASQC. 1994. *Quality systems—Model for quality assurance in production, installation, and servicing.* Milwaukee: ASQC. (This reference is the American version of the ISO 9002 standard.)

ASTM. 1990. *Manual on presentation of data and control chart analysis.* 6th ed. Philadelphia: American Society for Testing and Materials.

AT&T. 1985. *Statistical quality control handbook.* Indianapolis: AT&T Technologies.

Barrentine, Larry B. 1991. *Concepts for R&R studies.* Milwaukee: ASQC Quality Press.

Baumeister, T., E. Avallone, and T. Baumeister. 1978. *Marks' standard handbook for mechanical engineers.* 8th ed. New York: McGraw-Hill.

Bissell, A. F. 1990. How reliable is your capability index? *Applied Statistics* 39: 331–340.

Burns, Robert H., and Prabhat Kumar. 1996. Conquer corrosion in harsh environments with tantalum. *Chemical Engineering Progress* (March): 32–35.

Clausewitz, Carl von. [1831] 1976. *On war.* Book 1. Translated by M. Howard and P. Paret. Princeton, N.J.: Princeton University Press.

Contino, A. V. 1987. Improve plant performance via statistical process control. *Chemical Engineering,* 20 July, 95–102.

Cooper, P. J., and N. Demos, 1991. Losses cut 76 percent with control chart system. *Quality* (April).

Cryer, Jonathan D. 1986. *Times series analysis.* Boston: PWS-Kent Publishing.

Feigenbaum, Armand V. 1991. *Total quality control.* 3d ed. New York: McGraw-Hill.

Franklin, LeRoy A., and Gary S. Wasserman. 1992. A note on the conservative nature of the tables of lower confidence limits for C_{pk} with a suggested correction. *Communication in Statistical Simulations* 21, no. 4:1165–1169.

Harriott, Peter. 1964. *Process control.* New York: McGraw-Hill.

Holmes, Donald. 1988. *Introduction to SPC*. Schenectady, N.Y.: Stochos.

Holmes, Donald, and A. Erhan Mergen. 1989. Testing control chart subgroups for rationality. *Quality and Reliability Engineering International* 5:143–147.

Hradesky, John. 1988. *Productivity and quality improvement—A practical guide to implementing statistical process control*. New York: McGraw-Hill.

Johnson, Norman L., and Samuel Kotz. 1970. *Distributions in statistics: Continuous univariate distributions 2*. New York: John Wiley & Sons.

Juran, Joseph. 1995. *A history of managing for quality*. Milwaukee: ASQC Quality Press.

Juran, Joseph, and Frank Gryna. 1988. *Juran's quality control handbook*. 4th ed. New York: McGraw-Hill.

Kessler, Sheila. 1995. *Total quality service*. Milwaukee: ASQC Quality Press.

Kushler, Robert H., and Paul Hurley. 1992. Confidence bounds for capability indices. *Journal of Quality Technology* 24, no. 24:188–195.

Law, Averill M., and W. David Kelton. 1982. *Simulation modeling and analysis*. New York: McGraw-Hill.

Lawless, Jerald F. 1982. *Statistical models and methods for lifetime data*. New York: John Wiley & Sons.

Levinson, William. 1994. Multiple attribute control charts. *Quality* (December): 10–11.

———. 1996. Do you need a new gage? *Semiconductor International* (February): 113–117.

Mahan, Alfred Thayer. 1980. *The influence of sea power upon history. 1660–1805*. London: Bison Books.

Messina, William. 1987. *Statistical quality control of manufacturing managers*. New York: John Wiley & Sons.

Montgomery, Douglas. 1984. *Design and analysis of experiments*. 2d ed. New York: John Wiley & Sons.

———. 1991. *Introduction to statistical quality control*. 2d ed. Milwaukee: ASQC Quality Press and New York: John Wiley & Sons.

Moore, Albert W. 1994. A vital issue, a vital event. *Manufacturing Engineering* (August): 256.

Murphy, Robert, and William Levinson. 1996. Self-directed work teams. Paper presented at ASQC Annual Quality Congress, 14 May, at Chicago Hilton & Towers, Chicago, Illinois.

Musashi, Miyamoto. [1645] 1974. *A book of five rings.* Translated by Victor Harris. Woodstock, N.Y.: Overlook Press.

Nakajima, Seiichi. 1989. *TPM development program—implementing total productive maintenance.* Cambridge, Mass.: Productivity Press.

Peters, Tom. 1987. *Thriving on chaos.* New York: Harper & Row.

———. 1988. *Structures for the year 2000.* Palo Alto, Calif.: The Tom Peters Group.

Peters, Thomas, and Robert Waterman. 1982. *In search of excellence.* New York: Harper & Row.

Rose, E., R. Odom, and D. Pankey. 1996. Four key ideas for successfully implementing change focus: Transitioning to team-based management. Paper presented at ASQC Annual Quality Congress, 14 May, at Chicago Hilton & Towers, Chicago, Illinois.

Scotto, Michael J. 1996. Seven ways to make money from ISO 9000. *Quality Progress* (June): 39–41.

Shapiro, Samuel S. 1986. *How to test normality and other distributional assumptions.* Milwaukee: ASQC Quality Press.

Shirose, Kunio. 1992. *TPM for operators.* Cambridge, Mass.: Productivity Press.

Struebing, Laura. 1996. Customer loyalty: Playing for keeps. *Quality Progress* (February): 25.

Why companies fail quality audits. 1996. *Manufacturing Engineering* (May).

From the Internet/World Wide Web

(*Note:* These references were available as of June 1996. Since people can add or delete web pages anytime, there is no guarantee that they will be available in the future.)

Adam Smith (1723–1790). *An Inquiry into the Nature and Causes of the Wealth of Nations.* http://www.duke.edu/~atm2/index.html. Web page provided by Aaron Miller, a student at Duke University (as of 1996).

National Institute of Standards and Technology. Web page at http://ts.nist.gov/ts/htdocs/200.html

Thomas Paine. A supernumerary crisis. *Common Sense.* New York, December 9, 1783. Converted to HTML by Danny Barnhoorn for The American Revolution—an HTML project. http://ukanaix.cc.ukans.edu:80/carrie/docs/usdocs.txt/crisis13b.html

Index